NEBOSH NATIONAL DIPLOMA FOR OCCUPATIONAL HEALTH AND SAFETY MANAGEMENT PROFESSIONALS

UNIT ND3 - PART 1

ND3 Learning Outcome 11.1

ND3 Learning Outcome 11.2

ND3 Learning Outcome 11.3

ND3 Learning Outcome 11.4

ND3 Learning Outcome 11.5

ND3 Learning Outcome 11.6

ND3 Learning Outcome 11.7

CONTRIBUTORS

Peter Brookbank, MSc, Dip2.OSH, CMIOSH

For information on all RRC publications and training courses, visit: www.rrc.co.uk

RRC: ND3 - Part 1

ISBN for this volume: 978-1-912652-44-0
Fourth edition August 2022

ACKNOWLEDGMENTS

RRC International would like to thank the National Examination Board in Occupational Safety and Health (NEBOSH) for their co-operation in allowing us to reproduce extracts from their syllabus guides.

This publication contains public sector information published by the Health and Safety Executive and licensed under the Open Government Licence v.3 (www.nationalarchives.gov.uk/doc/open-government-licence/version/3).

Every effort has been made to trace copyright material and obtain permission to reproduce it. If there are any errors or omissions, RRC would welcome notification so that corrections may be incorporated in future reprints or editions of this material.

Introduction

ND3 Learning Outcome 11.1

ND3 Learning Outcome 11.2

ND3 Learning Outcome 11.3

ND3 Learning Outcome 11.4

ND3 Learning Outcome 11.5

ND3 Learning Outcome 11.6

ND3 Learning Outcome 11.7

Introduction

Course Structure

This study text has been designed to provide the learner with the core knowledge needed to successfully complete the NEBOSH National Diploma for Occupational Health and Safety Management Professionals, as well as providing a useful overview of health and safety management. It follows the structure and content of the NEBOSH syllabus and includes extra "Prior Learning" material to support your understanding of diploma-level content.

The NEBOSH National Diploma consists of three units of study. Learners must achieve a 'Pass' in all three units to achieve the qualification, and you need to pass the three units within a five-year period. For more detailed information about how the syllabus is structured, visit the NEBOSH website (www.nebosh.org.uk).

Unit 3 focuses on the controls of workplace safety issues and covers one Learning Outcome, divided into 12 sections.

Assessment Background

There are no in-person examinations for the National Diploma for Occupational Health and Safety Management Professionals. Instead, assessment will be via assignments and scenario-based case studies. Details of these together with sample assessments and assessment dates can be found on the NEBOSH website:

www.nebosh.org.uk/qualifications/national-diploma-for-safety-and-health-management-professionals/#assessments

NEBOSH state that the assessments are a substantial undertaking and should take around 60 hours for ND1, and 40 hours each for ND2 and ND3. It is important that you prepare well for the assessments and remember that whilst these may not be traditional exams, they are still intended to be challenging assessments of your capabilities and skills. You are therefore allowed a significant period of time to complete the assessment, so it is vital that you understand this and do not leave the assessment until the last minute.

Unit	Estimated time required to complete assessment	Time period given by NEBOSH to complete assessment
ND1	60 hours	6 weeks (30 working days)
ND2	40 hours	4 weeks (20 working days)
ND3	40 hours	4 weeks (20 working days)

During the assessments you will have access to books and the internet, however this must be your own work and there are stringent protocols in place to prevent plagiarism. All assessments will be checked via plagiarism software and each assessment also includes a closing interview. Guidance on digital assessments can be found on the NEBOSH website at:

www.nebosh.org.uk/digital-assessments/diploma

Unit ND3 Assessment

The Unit ND3 assessment consists of questions based on a fictitious but realistic scenario. The number of questions, the scenario and the tasks will change on each paper and will not cover the whole of the ND3 syllabus.

Results

Results will be issued 50 working days after the submission date for the assessment. After successful completion of each unit a 'unit certificate' will be awarded. After you have completed all 3 units the combined percentage mark will be used to determine your final grade:

- 226 or more: Distinction
- 196-225: Credit
- 150-195: Pass

More Information

As you work your way through this book, always remember to relate your own experiences in the workplace to the topics you study. An appreciation of the practical application and significance of health and safety will help you understand the topics.

Keeping Yourself Up-to-Date

The field of health and safety is constantly evolving and, as such, it will be necessary for you to keep up to date with changing legislation and best practice.

RRC International publishes updates to all its course materials via a quarterly e-newsletter (issued in February, May, August and November), which alerts students to key changes in legislation, best practice and other information pertinent to current courses.

Please visit https://www.rrc.co.uk/news-resources/newsletters.aspx to access these updates.

NEBOSH National Diploma for Occupational Health and Safety Management Professionals	
Unit ND1	Know – Workplace Health and Safety Principles (UK)
Unit ND2	Do – Controlling Workplace Health Issues (UK)
Unit ND3	Do – Controlling Workplace Safety Issues (UK)

Learning Outcome 11.1

NEBOSH National Diploma for Occupational Health and Safety Management Professionals

ASSESSMENT CRITERIA

- Outline the legal requirements and practical considerations for maintaining a safe working environment.

LEARNING OBJECTIVE

Once you've studied this Learning Outcome, you should be able to:

- Explain the need for, and factors involved in, the provision and maintenance of a safe working environment.

Contents

Safe Working Environment

IN THIS SECTION...

- The **Workplace (Health, Safety and Welfare) Regulations 1992** set out the general standards for health safety and welfare in the workplace to enable us to provide a safe working environment.
- A safe place of work includes the provision and maintenance of safe means of access and egress, and also cleanliness, workstations and seating, and windows and transparent doors.
- Safety signs are standardised so that they have the same meaning wherever they are used.
- The four categories of sign: prohibition, mandatory, safe condition and warning, are distinguished by their shape and colour and use symbols (pictograms) to convey their message.
- The **Health and Safety (Safety Signs and Signals) Regulations 1996** govern the use of signs and signals in the workplace.
- Employers have a duty under the **Management of Health and Safety at Work Regulations 1999 (MHSWR)** to ensure that lighting in the workplace is suitable and adequate. HSE Guidance Document HSG38 *Lighting at work*, gives best practice advice and guidance on this subject.

Workplace (Health, Safety and Welfare) Regulations 1992

The safe place of work is an important and general requirement of workplace health and safety, but it's difficult to specify absolutely. The **Workplace (Health, Safety and Welfare) Regulations 1992** and its accompanying **Approved Code of Practice** (ACoP) specify what is needed and in this section we will be examining the requirements under these regulations for a safe place of work, a clean workplace, safe workstations, safety windows or other transparent surfaces and general access and egress.

These regulations are accompanied by an ACoP which sets out practical considerations for health, safety and welfare in the workplace, enabling employers to provide a safe and healthy working environment. Below is a brief summary of the main topics covered.

- **Maintenance** of the workplace and of associated equipment in a good state of repair.
- **Ventilation** to provide a sufficient quantity of fresh or purified air.
- **Temperature** in indoor workplaces – reasonable during working hours (16°C or 13°C if the work involves severe physical effort).
  - Method of heating or cooling should not produce harmful fumes.
  - Thermometers provided to enable persons to determine the temperature.
- **Lighting** which is suitable and sufficient, natural light if possible and emergency lighting provided in the event of failure of artificial light.
- **Cleanliness** of workplace, furniture, furnishings and fittings and waste material not allowed to accumulate except in suitable receptacles.

- **Room dimensions** sufficient and not so **overcrowded** as to cause risk to the health or safety of persons at work in it. The HSE ACoP to the **Workplace (Health, Safety and Welfare) Regulations 1992** (L24), indicates that 11 cubic metres of space per person should be taken as the minimum and should not take in to account ceiling heights over 3m.
- **Workstations** suitable for any person at work at the workstation and any work that is likely to be carried out there. Suitable seating provided for sedentary work.
- **Floors** suitable for use, without holes or slopes, not uneven or slippery, kept free from obstructions likely to cause trips, slips or falls and with effective drainage.
- Measures to prevent persons **falling** a distance likely to cause personal injury, being **struck by a falling object** likely to cause personal injury, or **falling into a dangerous substance** in a tank, pit or structure (by secure cover or fencing).
- **Transparent or translucent surfaces** in windows, doors, gates, walls and partitions constructed of safety materials or adequately protected against breakage and marked to make their presence apparent.
- **Windows, skylights and ventilators** able to be opened, closed and adjusted safely without exposing persons to any risk to health and safety.
- Windows and skylights in a workplace designed and constructed so that they can be **cleaned safely**.
- **Workplace traffic routes** to allow for the segregation of pedestrians and vehicles and to allow both to circulate freely and safely.
- **Doors and gates** suitably constructed and fitted with safety devices where necessary.
- **Escalators and moving walkways** to function safely, be equipped with any necessary safety devices and fitted with one or more emergency stop controls.
- **Sanitary conveniences** and **washing facilities** provided which are adequately ventilated and lit, kept clean and in an orderly condition, with separate rooms provided for men and women.
- **Drinking water** provided for all persons at work.
- Accommodation for any person at work's own **clothing** which is not worn during working hours and for special clothing worn at work but which is not taken home.
- **Facilities for changing** where the person has to wear special clothing for the purpose of work.
- **Rest facilities** including for meals where food eaten in the workplace would otherwise be likely to become contaminated and for any person at work, who is pregnant or a nursing mother, to rest.

Workplace traffic routes should allow for the segregation of pedestrians and vehicles

Safe Places of Work, Safe Means of Access and Egress

Section 2(2)(d) of the **Health and Safety at Work, etc. Act 1974 (HSWA)**, places a duty on employers to provide a safe place of work, including access and egress. This means that the employer must take account of where the work is being carried out. There should be sufficient traffic routes, of sufficient width and headroom, to allow people on foot or in vehicles to circulate safely and without difficulty. Features which obstruct routes should be avoided. Consideration should be given to the safety of people with impaired or no sight.

HSWA also requires the employer to maintain the workplace in a condition that is safe and without risks to health. They must also provide and maintain means of access to and egress from it that are safe and without such risks.

The **Workplace (Health, Safety and Welfare) Regulations 1992**, and its accompanying Approved Code of Practice, set out in more detail what needs to be provided regarding specific issues such as ventilation, temperature, lighting, cleanliness, space, workstation design, floors, windows, traffic routes, doors, gates, escalators and moving walkways, and the standard to which they need to be maintained.

Workplaces, equipment, devices and systems must be provided and maintained in efficient working order and in good repair. So, if a potentially dangerous defect is discovered, the defect should be rectified immediately or steps should be taken to protect anyone who might be put at risk. A suitable system of maintenance involves ensuring that:

- Regular maintenance is carried out at suitable intervals (including, as necessary, inspection, testing, adjustment, lubrication and cleaning).
- Any potentially dangerous defects are remedied and access to defective equipment is prevented in the meantime.
- Regular maintenance and remedial work is carried out properly.
- A suitable record is kept to ensure the system is properly implemented and to assist in validating maintenance programmes.

Cleanliness

To achieve an acceptable standard of cleanliness:

- Workplaces and furniture, furnishings and fittings need to be kept clean.
- Surfaces of floor, walls and ceilings of workplaces in buildings should be able to be clean.
- Waste materials shouldn't be allowed to accumulate in the workplace, unless in suitable containers.

The standard of cleanliness required will depend on the use to which the workplace is put. For example, an area in which workers take meals would be expected to be cleaner than a factory floor, and a factory floor would be expected to be cleaner than an animal house.

Floors and indoor traffic routes should be kept as clean as practicable. In factories and other workplaces of a type where dirt and refuse accumulates, any dirt and refuse which is not in suitable receptacles should be removed at least daily. These tasks should be carried out more frequently, where necessary, to maintain a reasonable standard of cleanliness or to keep workplaces free of pests and decaying matter.

Workstations and Seating

Workstations should be arranged so that each task can be carried out safely and comfortably. The worker should be at a suitable height in relation to the work surface. Work materials and frequently used equipment or controls should be within easy reach, without the need for undue bending or stretching.

Unsuitable workstation and seating for display screen equipment use

Workstations including seating, and access to workstations, should be suitable for any special needs of the individual worker, including workers with disabilities.

Each workstation should allow any person who is likely to work there adequate freedom of movement and the ability to stand upright. Spells of work which unavoidably have to be carried out in cramped conditions should be kept as short as possible and there should be sufficient space nearby to relieve discomfort.

There should be sufficient clear and unobstructed space at each workstation to enable the work to be done safely. This should allow for the manoeuvring and positioning of materials, for example lengths of timber.

Seating provided should, where possible, provide adequate support for the lower back, and a footrest should be provided for any worker who cannot comfortably place his or her feet flat on the floor.

Windows and Transparent Doors

Windows or other transparent or translucent surfaces in a wall, partition, door or gate should be:

- Made of safety material or protected against breakage.
- Appropriately marked or incorporate features to make it apparent.

'Safety materials' are:

- Materials which are inherently robust, such as polycarbonates or glass blocks.
- Glass which, if it breaks, breaks safely.
- Ordinary annealed glass which meets the thickness criteria in the following table:

Thickness criteria of glass

Nominal Thickness	Maximum Size
8mm	1.10m × 1.10m
10mm	2.25m × 2.25m
12mm	3.00m × 3.00m
15mm	Any size

Provision of Safe Means of Access and Egress

There should be sufficient traffic routes, of sufficient width and headroom, to allow people on foot or in vehicles to circulate safely and without difficulty.

Floors and traffic routes should:

- Be of sound construction.
- Have adequate strength and stability, taking into account the loads placed on them and the traffic passing over them.
- Not be overloaded.
- Have no holes, slopes, or uneven or slippery surfaces likely to:
  - Cause a person to slip, trip or fall, or to drop or lose control of anything being lifted or carried.
  - Cause instability or loss of control of vehicles and/or their loads.
- Be kept free of obstructions which may present a hazard or impede access.

Open sides of staircases should be securely fenced and provided with a secure and substantial handrail.

Common Safety Signs and Their Categorisation

Prohibition Signs

- Are round with a white background and red border and cross bar.
- Have symbols in black placed centrally on the background without obliterating the cross bar.
- Meaning that something must not be done.

No smoking

Smoking and naked flames forbidden

Not drinkable

No access for unauthorised persons

No access for pedestrians

Do not extinguish with water

No access for industrial vehicles

Do not touch

Prohibition signs

Mandatory Signs

- Are round with a blue background and white symbol.
- State what must be done (or what protective equipment must be worn, as in the examples below).

Eye protection

Safety helmet

Protective footwear

Ear protection

Respirator

Hand protection

Mandatory signs

Safe Condition (Formerly Emergency) Signs

- Are square or oblong with white symbols on a green background.
- Indicate such safe conditions as a first-aid post or emergency evacuation route.

Exit

Exit

First aid

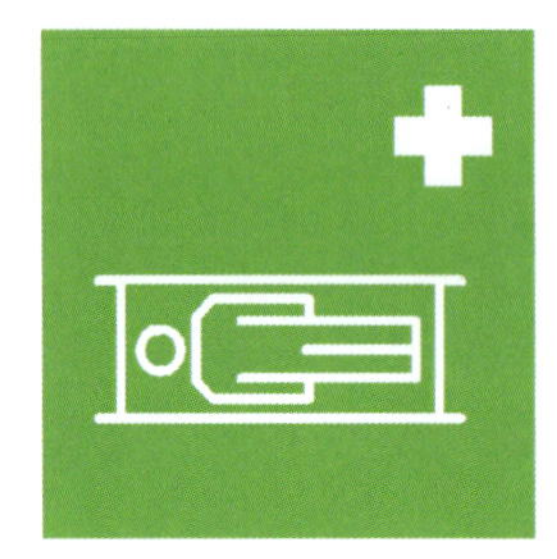

Stretcher

Eyewash

Safe condition signs

Warning Signs

- Are triangular with a black border and a black pictogram on a yellow background.
- Warn of the presence of a particular hazard.

Flammable material

Explosive

Industrial vehicles

Toxic

Corrosive

Electricity

Radioactive

Overhead loads

Unspecified general danger

Warning signs

Fire Safety Signs

Fall into two categories:

- Those providing information on means of escape and which take the form of a 'Safe Condition'.
- Those identifying the location of fire equipment, e.g. 'Alarm Point', 'Fire Extinguisher'.

A fire safety sign which bears only text such as 'Fire Exit', is not acceptable, although text may be used in combination with pictograms.

Acceptable fire safety sign

Fire equipment signs are square or rectangular in shape with a white pictogram on a red background.

Fire extinguisher

Fire hose

Fire equipment signs

Practical Considerations and Typical Areas Where Safety Signs Might be Used

Signboards should be:

- Installed in a position appropriate to the line of sight, either at the access point to the area of a general hazard or in the immediate vicinity of a specific hazard.
- Clearly visible in a well lit position.

Avoid placing too many signs close together and remove signs if the situation to which they refer no longer exists.

Illuminated signs should be provided with emergency lighting power if power might be lost in an emergency.

Pipework containing dangerous substances should be marked. In particular, it should be identified and marked at sampling and discharge points using the same symbols or pictograms as those commonly seen on containers of dangerous substances.

Corrosive material

Flammable material

Explosive material

Toxic material

Labelling pipework

Stores and areas containing significant quantities of dangerous substances should be identified by an appropriate warning sign unless the labels on the containers can be seen clearly from outside the store.

MORE...

Guidance is provided in BS 1710:2014 and BS 4800 on the use of different colours and safety signs to identify the contents of pipework and its associated risks.

Signs marking **obstacles**, **dangerous locations** and **traffic routes** can be used where:

- The risk is low.
- It is impractical to safeguard by other means.

Obstacles or dangerous locations such as the edge of a loading platform, or a danger zone adjacent to a process can be marked by the use of yellow and black (or red and white) angled stripes.

Where clearly defined **traffic routes** are necessary, they should be marked using continuous yellow or white lines.

Standard road traffic signs and markings should be used in outdoor areas to control vehicles and pedestrians.

Inspection pit with clearly marked edges

Health and Safety (Safety Signs and Signals) Regulations 1996

These regulations govern the use of signs and signals in the workplace; below is a very brief summary of the main issues:

- Aim to standardise safety signs so that wherever they are seen, they have the same meaning.
- Cover various means of communicating health and safety information including:
  - Traditional signboards, such as prohibition, warning signs and fire safety signs discussed previously.
  - Illuminated signs.
  - Hand and acoustic signals (e.g. fire alarms).
  - Spoken communication.
  - The marking of pipework containing dangerous substances.
- Require employers to:
  - Provide specific safety signs whenever there is a risk which has not been avoided or controlled by other means such as engineering controls and safe systems of work:
    - Where a safety sign would not help to reduce a risk, or where the risk is not significant, there is no need to provide a sign.
- Require, where necessary, the use of road traffic signs within workplaces to regulate road traffic.
- Require employers to:
  - Maintain the safety signs which they have provided.
  - Explain unfamiliar signs to their employees and tell them what to do when they see a safety sign.
- Apply to all places and activities where people are employed.
- Exclude signs and labels used in connection with the supply of substances, products and equipment or the transport of dangerous goods.

Workplace Lighting

Employers have a duty under the **Management of Health and Safety at Work Regulations 1999 (MHSW)** to ensure that lighting in the workplace is suitable and adequate and which also meets the provisions of the **Workplace (Health, Safety and Welfare) Regulations 1992**.

MORE...

Assistance in managing the health and safety risks from lighting in the workplace can be found in the HSE Guidance document HSG38 *Lighting at work* on the HSE website:

www.hse.gov.uk/pubns/priced/hsg38.pdf

Good lighting at work is important and regardless of the legal requirements, contributes to increased employee morale and better work ambience leading to improved workplace performance.

Poor lighting in the workplace can contribute to accidents (where employees cannot see hazards because of low lighting levels) and ill health including, stress, eyestrain, migraine, headaches, and has also been recognised as an aspect of sick building syndrome.

The cost to the employer becomes evident when employees have to take time off as a result of accidents or ill health with the consequent reduced productivity and loss of orders.

Managing the health and safety risks from lighting in the workplace includes:

- **Planning** - where employers need to identify priorities and set targets for improvement by assessing whether the lighting is suitable and safe for the work being carried out in the workplace.
- **Organising** - employees should be made aware of their responsibilities to look after themselves and others who may be affected by their actions, and if lighting is proving to be a problem in the workplace then they should bring this to the attention of the employer. The employer should also ensure there are competent maintenance staff available to deal with any lighting problems.
- **Control** - lighting should be maintained in an efficient state, in efficient working order and in good repair which means checking on a regular basis and ensuring the lighting levels are maintained.
- **Monitoring** - HSG38 gives a practical indication of the illuminance level requirements for the workplace (measured illuminance in lux) as set out in the Chartered Institution of Building Services Engineers (CIBSE) *Code for Lighting*. Monitoring lighting conditions will ensure these standards have been met.
- **Risk assessment** - when assessing lighting in the workplace, employers should consider whether the arrangements for workplace lighting are satisfactory or might present a risk to employees, contractors or visitors to the premises. In doing so, employers must take into account hazards such as glare, colour effects, stroboscopic effects and flicker. Other issues might include 'veiling effects' (where high luminance reflections overlay the detail of a task) and visible radiation from infrared and ultraviolet radiation from tungsten halogen lamps, high-intensity discharge lamps, carbon-arc and short-arc lamps and display lasers. High-power lamps used in theatres, broadcasting studios and entertainment also present a risk as they require very high output lighting for filming and performance work. The risk assessment itself should follow the same five-step principle of:
  - Look for the hazards.
  - Decide who might be harmed and how.
  - Evaluate the risks.
  - Record the findings.
  - Review the assessment regularly.

TOPIC FOCUS

Assessing Lighting in the Workplace

"It is important that lighting in the workplace:

- *allows people to notice hazards and assess risks;*
- *is suitable for the environment and the type of work (for example, it is not located against surfaces or materials that may be flammable);*
- *provides sufficient light (illuminance on the task);*
- *allows people to see properly and discriminate between colours, to promote safety;*
- *does not cause glare, flicker or stroboscopic effects;*
- *avoids the effects of veiling reflections;*
- *does not result in excessive differences in illuminance within an area or between adjacent areas;*
- *is suitable to meet the special needs of individuals;*
- *does not pose a health and safety risk itself;*
- *is suitably positioned so that it may be properly maintained or replaced, and disposed of to ensure safety;*
- *includes, when necessary, suitable and safe emergency lighting."*

(Source: HSG38 *Lighting at work*)

Good Practice for Workplace Lighting

HSG38 also gives guidance on good practice when it comes to designing and installing lighting systems in the workplace.

Different workplaces will require differently designed lighting systems and there is no 'one size fits all' when it comes to lighting levels and luminescence. Small, indoor workshops carrying out close contact work of a difficult nature will have different requirements to large, outdoor workplaces or even those working inside large, open-plan buildings. So the type of work being carried out and the working environment are important factors in deciding on the design of the lighting system.

Another aspect of the working environment is - what levels of natural light are available and is it possible to maximise the use of it (most people prefer to work in natural daylight) or does this have to be augmented with artificial lighting? Some workplaces may have no access to natural lighting and must have full artificial lighting which will itself bring problems ensuring the lighting design is suitable for workers so the design avoids glare and the effects of veiling reflections. Lighting system designers should try to avoid dark and/or reflective working surfaces. Light reflected off the walls is usually distributed more evenly than direct lighting and can soften shadows and reduce the effects of any veiling reflections and glare.

Working conditions are important - particularly in areas that might create a dusty, flammable or explosive atmosphere. Lighting design should reflect the local environment and ensure that objects that may be combustible are not damaged by lamps that operate at high temperatures. There is a CIBSE Guidance document, *Lighting in hostile and hazardous environments* and the provisions of the **RRFSO** (Fire Risk Assessment) and Regulation 5 of **DSEAR** should also be taken into account when designing and installing lighting systems in these environments. **DSEAR** identifies 'zones' which relate to the presence, or possible presence, of flammable atmospheres and the **Equipment and Protective Systems Intended for Use in Potentially Explosive Atmospheres Regulations 2016** legislate on suitable electrical equipment that may be used in those zones.

The employer has a duty to ensure the health and well-being of employees and individual requirements have to be considered when designing and installing lighting systems. Workers who have to get closer to their screens or workpieces in order to see what they are doing, are likely to suffer eyestrain and adopt unsuitable postures leading to musculoskeletal problems. Vulnerable individuals such as those with epilepsy may have to be provided with appropriate lighting and protected from equipment with a flickering light as this can trigger a seizure.

The employer should have a safe and environmentally robust system in place for the maintenance, replacement and disposal of lamps and luminaires. Ease of access to remote or high level luminaires such as having the use of a Mobile Elevated Work Platform (MEWP) or being able to wind the lamp unit down to ground level for maintenance, is preferred to asking maintenance personnel to work off tall ladders or having the disruption of installing scaffolding in the workplace.

STUDY QUESTIONS

1. Under the **Health and Safety (Safety Signs and Signals) Regulations 1996**, when is an employer required to provide safety signs?
2. Describe the shape and colour of: prohibition, mandatory, safe condition, and warning signs.
3. What should employers consider when installing lighting at their workplace?

(Suggested Answers are at the end.)

Summary

Safe Working Environment

We have:

- Explained the role of the **Workplace (Health, Safety and Welfare) Regulations 1992** in ensuring workplaces are safe.
- Outlined what constitutes a safe working environment, including design and maintenance issues.
- Identified the common types of safety sign - prohibition, mandatory, safe condition and warning signs - displayed in the workplace.
- Outlined where signs should be used and considered how other standard warning markings and signs should be used.
- Explained the role of the **Health and Safety (Safety Signs and Signals) Regulations 1996**.
- Outlined the requirements for lighting provision in the workplace and good practice on designing and installing workplace lighting systems as set out in HSE Guidance Document HSG38 *Lighting at work'*.

Learning Outcome 11.2

NEBOSH National Diploma for Occupational Health and Safety Management Professionals

ASSESSMENT CRITERIA

- Recognise risks and design safe working practices in confined spaces.

LEARNING OBJECTIVE

Once you've studied this Learning Outcome, you should be able to:

- Explain the hazards, risks and control measures associated with work in confined spaces.

Contents

Confined Spaces

IN THIS SECTION...

- The **Confined Spaces Regulations 1997** lay down the legal requirements for work in confined spaces.
- Examples of where confined space entry may occur in the workplace include entry into inspection pits in garages, trunking ducts, watercourses, trenches, silos and sewers.
- In assessing the risk from working in a confined space we need to consider:
  - Available means of access/egress.
  - Atmospheres that are likely to be encountered.
  - Tasks, materials and equipment to be used.
  - The persons at risk.
  - The reliability of safeguards in place.
- Employers must provide a safe system of work, and this is vital when carrying out confined space entry and work so safe working practices must include:
  - Permit-to-work systems.
  - Emergency arrangements.
- Effective training for work in confined spaces is necessary for all workers.

Meaning of the Term 'Confined Space'

The **Confined Spaces Regulations 1997** define a confined space as *"any place, including any chamber, tank, vat, silo, pit, trench, pipe, sewer, flue, well or other similar space in which,* ***by virtue of its enclosed nature****, there arises a reasonably foreseeable* ***specified risk****"*.

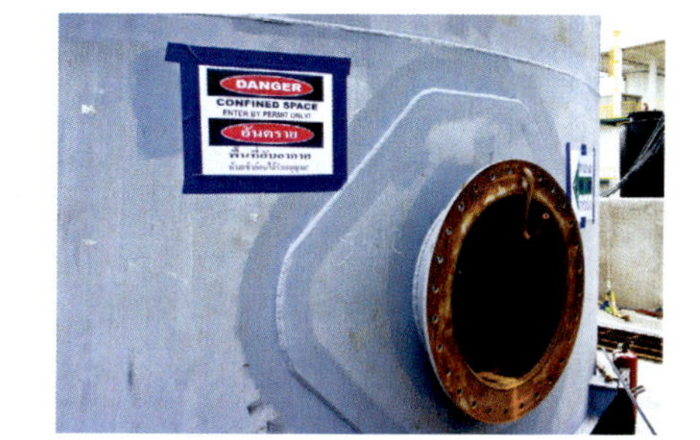

Working in a confined space is a high-risk activity

A "specified risk" means a risk of:

- Serious injury to any person at work arising from a **fire or explosion**.
- The loss of consciousness of any person at work arising from an **increase in body temperature**.
- The loss of consciousness or asphyxiation of any person at work arising from **gas**, **fume**, **vapour** or **lack of oxygen**.
- The **drowning** of any person at work arising from an increase in the level of a liquid.
- The **asphyxiation** of any person at work **arising from a free-flowing solid** or the inability to reach a respirable environment due to **entrapment** by a free-flowing solid.

DEFINITION

FREE-FLOWING SOLID

Any substance consisting of solid particles and which is of, or is capable of being in, a flowing or running consistency, and includes flour, grain, sugar, sand and other similar materials.

Confined spaces are recognised as workplaces that can be particularly hazardous, such as pits in garages, trunking ducts, watercourses, trenches, tanks, silos and sewers. When things go wrong inside a confined space they can often go dramatically wrong very quickly.

Confined space accidents often result in fatality, so the accident triangle (covered in Unit 1 Learning Outcome 5) for these types of accidents has quite steep sides (there are a lot of fatalities relative to the other types of injury that might result). In addition, confined space accidents often result in multiple casualties/fatalities (either because there was more than one worker in the space at the time of the incident, or because an attempt was made to rescue a casualty and the would-be rescuer themselves became a casualty).

Two farmers drowned in December 2015 after entering a slurry pit to remove a blockage. According to the pathologist, the farmers entered the pit and would have been overcome with toxic slurry fumes within a matter of seconds. (*FarmingUK*, 18 May 2018.)

The **Confined Spaces Regulations 1997** lay down the legal requirements for work in confined spaces. In simple terms, the law requires that no person at work shall enter a **confined space** to carry out work for any purpose unless:

- It is not reasonably practicable to achieve that purpose without such entry.
- In accordance with a **system of work** which, in relation to any relevant **specified risks**, renders that work safe and without risks to health.
- There have been prepared, in respect of that confined space, **suitable and sufficient arrangements** for the rescue of persons in the event of an emergency, whether or not arising out of a **specified risk**.

Examples of Confined Spaces in the Workplace

Confined spaces are defined by the possibility of a foreseeable **specified risk**. The types of specified risk can give us some indication of what might constitute a confined space in a work situation. So, for example:

Slurry pit

- **Fire or explosion** - e.g. a storage tank previously used to store a highly-flammable liquid; a new storage tank that has been sprayed with a lining using a flammable paint solvent; a tank that has been cleaned with a highly-flammable solvent cleaner.
- **Inhalation of gas, fumes, vapour or lack of oxygen** - e.g. a sewer where hydrogen sulphide (H_2S) levels may reach toxic concentrations; a ship's hold where rusting on the metal of the hull may have depleted the oxygen concentration below 17%; a storage tank containing oil residue that releases volatile vapours when disturbed; a vehicle inspection pit in a vehicle maintenance garage that may fill with exhaust fumes; trunking ducts carrying toxic vapour.
- **Drowning** - e.g. a deep excavation with a high-pressure water main pipe that if damaged would allow water to rapidly flood the excavation.
- **Entrapment in a free-flowing solid** - e.g. a grain silo containing wheat or barley where a worker standing on the top of the grain may fall into hidden voids created by the grain 'bridging' inside the silo.

- **Heat stress** - e.g. a steel storage vessel during a welding operation; inside an industrial boiler if sufficient cool-down time has not been allowed.

It is important to recognise that many places have an enclosed nature, but not all are confined spaces. For example, a broom cupboard under the stairs has an enclosed nature, but it is not a confined space because it does not present one or more of the five specified risks identified above. Just because a space is constricted does not make it a confined space.

Similarly, some spaces are enormous; there are no space constraints inside them. But if that space has an enclosed nature and it presents one or more of the specific risks then it is a confined space. So, for example, a 1,000,000L water tank is very large inside, but it does have an enclosed nature and if it is connected to the mains water supply then there is a risk of drowning as a result of increase in liquid level. Therefore, it is a confined space.

Factors to be Considered When Assessing Risk

Having studied the examples where confined spaces might occur in the workplace, we shall now look at some of the factors that should be considered when assessing the risk of work involving confined space entry.

Access Arrangements

Confined space entry

No-one may enter or remain in a confined space unless under a Safe System of Work (SSW). This may mean:

- The worker is wearing approved breathing apparatus if dangerous fumes are present or in an oxygen-deficient atmosphere.
- The worker has been authorised to enter by a responsible person.
- Where practicable, the worker is wearing a belt or harness with a retrieval line securely attached.
- A person is keeping watch outside.

- The other end of the retrieval line must be attached to a mechanical device or a fixed point outside the permit space. A mechanical device must be available to retrieve someone from vertical-type permit spaces more than five feet (1.52m) deep.

Alternatively, a person may enter or work in a confined space without breathing apparatus provided that:

- Effective steps have been taken to avoid ingress of dangerous fumes.
- Sludge or other deposits liable to give off dangerous fumes have been removed.
- The space contains no other material liable to give off such fumes.
- The space has been adequately ventilated and tested for fumes.
- There is a supply of air adequate for respiration.
- The space has been certified by a responsible person as being safe for entry for a specified period without breathing apparatus.

The routes into and out of a confined space must be considered. In some cases, large entrance ways at ground level will be available and these give relatively easy and unrestricted access into confined spaces. In other cases, however, entry into a confined space may be more difficult, such as climbing down a ladder. This may be a ladder temporarily placed into the confined space (e.g. a ladder placed into a 4m deep excavation) or permanently fixed in place (e.g. a vertical ladder fixed to the side of a ship's hold). Frequently, entry into the confined space will be through an opening, hatch or manway that is relatively constricted in size/diameter and that may present a significant obstruction to the passage of people and equipment/material. Sometimes there are multiple complicating factors to take into account. For example, entry into a sewer may be through a 5m deep inspection chamber of restricted dimensions, down a fixed vertical access ladder and then horizontally along the sewer line itself which is of limited width and height.

Likely Atmospheres to be Encountered

It is standard practice to 'purge' pipelines and tanks with inert gas, normally nitrogen to remove contaminants prior to entry. In the case of sewers, it may be enough to lift manhole lids along the line of work to allow for lighter than air gases to escape. Once the contaminant has been removed, the atmosphere must remain safe. It is still a requirement to continually monitor the atmosphere whilst work is in progress

A critical factor in many confined space entry assessments is the makeup of the atmosphere inside the space.

One key issue is the oxygen content of the atmosphere. Normal air has an oxygen concentration of 21%. If oxygen concentrations fall significantly below this concentration then there is a risk that workers inside the space will suffer the effects of breathing a reduced-oxygen atmosphere - known as 'hypoxia'. Symptoms may be mild - fatigue, confusion, nausea; to severe - loss of consciousness, loss of blood pressure, death.

Conversely, some confined spaces may have an increased oxygen content (e.g. caused by the leakage of oxygen from storage cylinders left inside the confined space). Whilst this will not cause any severe medical issues for the workers, it does significantly increase the fire/explosion risk inside the space as flammable gases/vapours will ignite more easily when oxygen concentrations are high (minimum ignition energy is decreased), and the subsequent fire will burn faster and consequently release more energy. So, fires are easier to start and more violent in high-oxygen concentration atmospheres.

The other critical factor that must be considered regarding the atmosphere inside the space is the presence of any other contaminant such as a gas, vapour or fume that might:

- affect workers directly by causing them to lose consciousness/asphyxiate; or
- increase the fire/explosion risk.

With oxygen making up 21% of the atmosphere, the rest is mostly made up of nitrogen (78%), a non-toxic, inert gas; however, other gases that may be present in the confined space may not be so benign. Many gases have a direct toxic effect on the body, such as carbon monoxide (CO), carbon dioxide (CO_2), hydrogen sulphide (H_2S) and chlorine (Cl_2). Some gases are flammable at relatively low concentrations and so can create an explosive atmosphere, such as methane (CH_4), butane (C_4H_{10}), propane (C_3H_8) and acetylene (C_2H_2). Even completely inert gases, such as non-flammable and non-toxic argon, may drive down the oxygen concentration to a point where hypoxia results.

It must be recognised that the atmosphere that exists inside a confined space may not stay constant during the work period. Oxygen levels may deplete as a result of respiration by the workers inside the space, or hot works taking place inside the space. Similarly, oxygen levels may increase as a result of leaks/releases from oxygen sources inside the space. Gas, vapour and fume concentrations may also vary over time. For example, hydrogen sulphide gas concentrations may be low at the start of work in a sewer, but may rise to dangerous levels within just a few minutes of entry as a result of the disturbance of materials in the sewer. The likelihood of changes in atmosphere must be factored into the risk assessment process.

Task, Materials and Equipment

The confined space risk assessment should consider whether the **tasks** to be carried out will increase the risks to the persons working in the confined space. Welding, for example, could reduce the amount of oxygen in the air and may also lead to heat stress. Strenuous physical activity will lead to heat stress and may also deplete oxygen concentrations if ventilation of the space is restricted. Inevitably, some tasks will carry inherent risks that have nothing to do with the fact that the task is taking place inside a confined space; the risks arise from the task itself. For example, working at height inside the space will have inherent risks that have nothing to do with the five specified risks but must still be factored into the risk assessment.

It is also necessary to consider the **materials** that will be used during the work activity. For example, cleaning with solvents or carrying out painting operations could produce an explosive atmosphere; the use of powders could produce large amounts of dust that might produce an explosive atmosphere or interfere with breathing. This is less to do with the materials that already exist within the confined space and more to do with the materials that are taken into the confined space in order for work to be carried out (e.g. an oxy-acetylene cutting set) or materials that are created by the work activity itself (e.g. welding fumes).

Any work **equipment** being used should also be part of the risk assessment. For example, trailing leads from electrical equipment can increase the chance of trips; powered hand tools can create a range of mechanical hazards (e.g. contact with a cutting disc) and non-mechanical hazards (e.g. electricity, noise, vibration, etc.). The equipment taken into the confined space must be carefully considered during the risk assessment to ensure that it does not itself create or increase one of the specific risks (e.g. a non-intrinsically safe communications set taken into a flammable atmosphere) and to ensure that the risks inherent with the use of the equipment (which have nothing to do with the confined space) are correctly addressed.

People at Risk

When assessing the risk associated with entry into, and work being carried out in, a confined space, it is necessary to consider who may be put at risk by the operation:

- How many people will be entering into the confined space?
- How many people will be involved in emergency rescue should that be required?
- Are workers medically fit to undertake the required work inside the space and are they medically fit to wear the necessary PPE, such as Breathing Apparatus (BA), that might be required for normal use or for emergency evacuation?

- Are workers competent to undertake the work? Competent means trained, skilled, experienced and with the right knowledge and ability to be able to work safely.
- Are those involved in emergency response medically fit and competent to undertake their duties?

Reliability of Safeguards (Including Personal Protective Equipment)

One final factor to take into account in the risk assessment process is the reliability of the safeguards chosen. As you may know from previous studies, control measures that rely on personal behaviour are often the least reliable whilst those that eliminate the hazard or address the hazard directly tend to be more reliable. PPE is usually the least preferable option because it is the least reliable over time, its use is very susceptible to human error and wilful misuse, and if PPE fails, it always fails to danger.

Controls and safeguards should be in place

So, during the risk assessment process, it is important to consider the range of options that might be available and the effectiveness and efficiency of those control options. Steam-cleaning a tank may seem like a difficult option if in-situ equipment does not exist, but it is preferable to putting workers into a tank to scrape up residue by shovel and bucket. The hierarchy of risk control options can be useful during this thinking process.

The other issue to take account of is that when choosing a specific type of control measure to use, there will always be a range of options available and those may vary in their reliability. For example, if BA is to be used to gain entry into a storage tank, there are lots of different types to choose from. There are lots of manufacturers of BA and, even though their products can be very similar in general specification, there may still be significant design and quality variation that might make one set more reliable compared to another. It is important to recognise that sometimes a standard item of PPE will be good enough for the job; as long as it meets the general specification for that item of PPE then the exact make and model do not matter. But for other safety critical items of PPE, the exact make and model will make a difference and therefore the most reliable item must be successfully identified and sourced.

Safe System of Working Practices

Operating Procedures

Since confined space entry is an inherently high-risk activity, a high level of management control must always be exercised. The following principles must be applied:

- Do not enter the confined space for any reason if it is possible/reasonably practicable to do the work in some other way.
- If it is not possible/reasonably practicable to do the work without gaining entry to the confined space then a competent person must undertake a thorough risk assessment of the work.
- A comprehensive SSW must be developed for the confined space entry.
- This SSW must incorporate the use of a permit-to-enter system to control access into the space.
- Emergency arrangements must be developed as a part of the SSW and personnel and equipment must be provided to carry out the emergency response.
- Everyone involved in the confined space entry work, including emergency responders, must be trained and competent in their roles.

Need for Access

There must be a justification of the need to enter the confined space. Consideration must be given to not doing the work at all. If this is not an option then consideration must be given to dismantling the confined space so that it no longer has an enclosed nature (e.g. a bolted together powder storage bin in a factory environment might be dismantled so it no longer has a top and sides). Further consideration must be given to the possibility of undertaking the work without putting workers into the space. For example, a silo can be cleaned by lowering a mechanical flail, called a **bin whip**, into the silo to sweep the sides clean. This negates the need to put workers into the silo. Technological advances such as sewer and pipeline inspection robots mean that there are often a range of alternatives to worker entry. The reasonably practicability of these options must be taken into account before a decision is made to put workers into the confined space.

Operation of a Permit-to-Enter System

The only possible way to ensure that proper management control is exercised over confined space entry is by using a permit to work, so this must be a **mandatory** requirement. This is often referred to as a 'permit to enter'. The permit system must be under the control of a **competent manager** and will detail the operating procedures to be adopted.

Operation of an effective permit-to-enter system is only possible if there is a clear understanding by all workers of what spaces constitute confined spaces and that these confined spaces are subject to permit-to-work control. Therefore, it is imperative that any workplace where there may be confined spaces undertakes a strategic review across their premises to identify all possible confined spaces and that these are then **clearly identified** for all workers who might potentially need to gain entry. This is best done by clearly labelling all confined spaces with appropriate signs and by incorporating a basic understanding of confined space policy as a part of site induction training for both direct workers and contractors.

The first and most important step is the assessment of the work by a **responsible person** familiar with the technical aspects of the job. This person must be competent - meaning trained, knowledgeable and experienced. They must be given time to consider the job in detail, so their workload must not be too heavy.

A typical decision sequence is shown in the following figure:

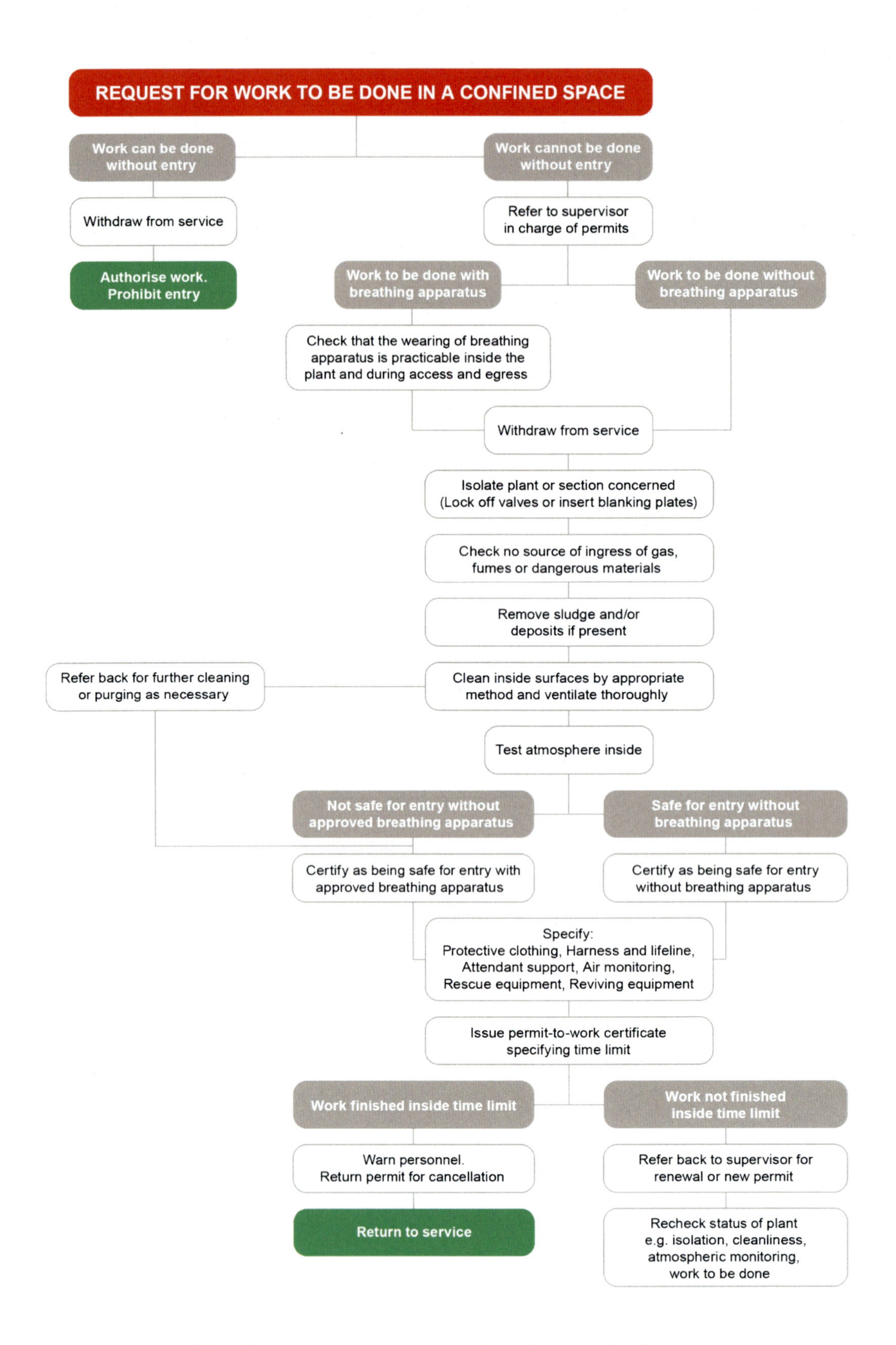

Typical arrangements associated with work in a confined space

The exact operation of the permit to work will vary depending on the workplace and type of work activity, but typical steps would include:

- **Issue**

  The issuing of the permit to the work group who are to undertake the entry. This would normally incorporate details of the work to be undertaken, identification of the space, the names of the work group involved, the specific date and time period over which the permit will remain open, the specific precautions that are required to ensure safety during the work, restrictions and reference to other permits that may also apply to the work.

- **Receipt**

  The formal handover of the space from the authorising manager to the work group undertaking the entry; incorporating names, signatures, dates and times to formalise the handover process.

- **Clearance/Return to Service**

  The formal return of the space from the work group back to the authorising manager. This will often incorporate confirmation that the space has been returned back to a safe condition, that all workers are out and that the space can be de-isolated and returned back to routine service.

  Again, this step will often incorporate signatures, dates and times to properly formalise the hand-back process.

- **Cancellation**

  Once all work has been completed and the confined space is fit to return back to normal service then the permit must be formally cancelled to render it null and void. This prevents the permit from remaining open and the potential for re-entry by members of the work group.

Permit-to-Work System

It is important to remember that a permit to enter is just a piece of paper. It is only as good as the management system that it forms a part of. Therefore, this permit-to-work system will only be effective in operation if it is properly adhered to by management and workers. This requires that the permit-to-work system is treated as a critical management control and not simply a piece of bureaucracy. The permit system must be checked and audited as a matter of routine to ensure that it remains robust.

Factors to be Considered when Developing the Safe System of Work

When developing the SSW for confined space entry, the competent person will have to consider a range of factors. These will vary depending on the nature of the confined space and the nature of the work undertaken. However, the following general issues must be considered:

Level of Supervision

All confined space entry work must be supervised to some degree. The exact level of supervision will vary depending on the inherent risks associated with the work, the complexity of the task and the competence of the workers undertaking the work. It will be entirely appropriate for some confined space entry work to be under constant supervision whilst other work requires a lower level of oversight.

Lone workers should never undertake confined space entry. There should always be a worker outside the confined space to raise the alarm and render help as necessary in the event of an incident. This person is commonly referred to as the **top man**. They may or may not have a role in the supervision of the work.

Competency Requirements of the Workers

Workers who undertake work inside confined spaces must be competent, but there is no one single standard for competence in this respect. Risks associated with confined spaces will vary depending on the space, and the nature of the task will also vary, so it is necessary to identify the specific competency requirements as a part of the risk assessment/permit-to-work procedure. For example, some confined space entry will require workers to wear BA. If this is the case then they must be trained and competent in the use of that BA. Other confined space entry will not require workers to have any BA and so this competence requirement is not relevant.

Communication Methods

Workers inside a confined space must be able to communicate with people (such as the top man) outside. In some instances, communication is very easy because the people involved are very close to one another and can talk to each other unimpeded. In other cases, special communication equipment such as handheld VHF radio sets must be used. Any communication equipment must be carefully selected so that it is effective (e.g. VHF radios only work within line of sight and do not transmit through dense material such as soil or rock) and safe (e.g. a VHF radio set may be an ignition source in a flammable atmosphere and so must be intrinsically safe).

Ventilation and Atmospheric Testing

The foreseeable atmospheres inside the confined space must be identified and the space may then have to be ventilated before entry is gained to ensure the atmosphere has sufficient oxygen and is free of contaminants (that might be flammable and/or toxic).

This might be done by:

- relying on **natural ventilation** where the space is opened to the atmosphere for some time prior to entry (e.g. opening of hatches 24 hours prior to the work being undertaken); or
- **forced air ventilation** where outside air is pumped into and through the space for a period of time (e.g. for the 24 hours prior to entry). This may involve the use of:
  - Compressed air.
  - Blower fan and trunking.
  - Exhaust fan or ejector and trunking.

It may be necessary to test the atmosphere prior to entry. For many confined spaces this testing is done once prior to entry and does not have to be repeated because there is no likelihood of any contaminant escaping into the space or oxygen levels dropping to dangerous levels.

In some cases, routine testing will be undertaken prior to entry and also during the work activity itself at a frequency identified by the competent person.

In other instances, the workers inside the confined space will carry portable personal gas detectors/alarms with them so that the levels of oxygen and other gases are constantly being monitored and alarms given if oxygen levels fall or other gas concentrations start to rise towards dangerous levels.

In many industries, the concentration of flammable or explosive gases should not be allowed to rise above 25% of the Lower Explosive Limit (LEL) but sometimes (particularly in the oil and gas industry) the level is set at 0% of the LEL. (See Learning Outcome 11.3 for more on lower explosive limits.)

Removal of Residues

If the confined space contains residues of previous contents (e.g. sludge left at the bottom of an oil storage tank) then it may be necessary to remove those residues before entry is gained. This may be done by steam cleaning or a chemical cleaning process.

Cleaning methods include:

- Steam.
- Partial filling with water and boiling.
- Washing with hot or cold water.
- Use of solvents or neutralising agents.

On completion of cleaning, flammable vapours may be **purged** with an inert gas.

If residues have to be removed by hand before other work can be undertaken then due consideration must be given to the risk that may be created. For example, vapour may be given off from the residue as it is removed, giving rise to a flammable asphyxiating atmosphere.

Isolation and Lock-Off

The various infeeds and outfeeds that allow contents into and out of the confined space must be identified and isolated. For example, the mains water supply to a water storage tank must be correctly identified and the valve turned off. Isolation must be done by the most effective means possible and must be secured. This means that valves should be padlocked in the closed position to ensure that it is not possible for them to be accidentally opened.

The same isolation and lock off of the various services that supply the confined space must also be achieved. For example, the electrical supply to the motor driving an agitator inside a tank must be securely isolated. Other services such as pneumatic, hydraulic, gas and steam must be similarly identified, isolated and secured.

In some cases, blanking plates may be inserted into pipelines and sections of pipeline removed to guarantee that their contents cannot enter the confined space.

TOPIC FOCUS

Safe Isolation Steps

1. Plant that is to be entered with power isolated is to be stopped by normal means.
2. All residual energy reserves, be they pneumatic, hydraulic, electric, etc. must be exhausted or discharged.
3. Liquid or free-flowing solid lines should be positively isolated using flange blanks or blinds.
4. Steam lines, because of the danger they represent, are normally isolated using 'double block and bleed', which involves the use of a three-valve system where a pipe has two closed valves and an open drain valve positioned between them so that material is prevented from flowing.
5. All moving parts must be stopped in a safe position that is suitable for the work to be carried out.
6. The electrical main isolator is turned **off**. This is the primary means of isolation on most plant and equipment.
7. A padlock is fitted to the isolator to secure it in the off position. Locks should ideally be labelled or coded to identify the owner of the lock, the date and time of isolation and permit reference.

Emergency stop buttons with integral locks normally only lock out the control circuitry and therefore are not suitable for access into confined spaces. Access should only be carried out with the main isolator locked off.

Personal Protective Equipment

There is no one single type or set of PPE requirements when undertaking confined space entry. It may or may not be necessary for workers to wear:

- Hard hats.
- Eye protection.
- Respiratory protective equipment such as respirators or breathing apparatus.
- Gloves.
- Body protection.
- Protective footwear.

The competent person will have to specify the PPE requirement during their risk assessment/permit-to-work issue process and there must then be an adequate level of supervision to ensure that the mandatory PPE is worn by all.

Breathing Apparatus

Work may be carried out in a confined space without BA if steps have been taken to ensure a breathable atmosphere. This means:

- Effective steps have been taken to avoid ingress of dangerous fumes.
- Sludge or other deposits liable to give off dangerous fumes have been removed.
- The space contains no other material liable to give off such fumes.
- The space has been adequately ventilated and tested for fumes.
- There is a supply of air adequate for respiration.
- The space has been certified by a responsible person as being safe for entry for a specified period without BA.

If, however, the use of BA is necessary, the following guidance should be followed:

- BA sets must generally be of a type approved by the enforcement authority.
- BA sets must be well-fitting and properly worn.
- A positive pressure inside the mask is recommended.
- Respirators must **not** be used in an oxygen-deficient atmosphere (as they simply filter the surrounding air).
- A safety harness and lifeline should be worn, with the free end held by a person outside.

Fresh air hose breathing apparatus

Source: HSG53 *Respiratory protective equipment at work - A practical guide*, HSE, 2013 (www.hse.gov.uk/pubns/priced/HSG53.pdf)

MORE...

The current applicable standards for breathing apparatus are: BS EN 14593 (airline) and BS EN 137 (self-contained).

Further information can be obtained from the HSE publication, HSG53 *Respiratory protective equipment at work - A practical guide*, available at:

www.hse.gov.uk/pubns/books/hsg53.htm

Access and Egress

Appropriate safe means of access must be identified as a part of the SSW. As mentioned previously, in some cases confined spaces can be entered very easily by opening a hatch or doorway and walking through.

In other cases, entry will have to be gained by ladder. If workers are encumbered by wearing items of PPE (such as breathing apparatus) then it may be very difficult to safely access the space and due consideration must be given to the risk of falls. For this reason, a hoist or winch system may be used as the safe means of access into the space, or as a safety standby.

Consideration must also be given to the fact that workers may need to leave the confined space quickly in the event of certain foreseeable emergencies, such as fire, in an excavation. Therefore, an appropriate number of emergency escape routes must be provided and, where possible, these must be near to the workers' positions in the space. For example, in a long excavation that has been identified as a confined space and where a significant risk of fire exists, multiple ladders might be positioned in the excavation with short travel distances from the workers to the base of each ladder.

The fact that workers may be wearing bulky items of PPE must be taken into account when specifying safe access. And to add to the complications, the fact that casualties may need to be evacuated from the confined space must also be factored in. A worker on a stretcher cannot be carried safely up a ladder and may not fit through a hatch or manway. Provision must be made for safe evacuation of casualties in the event of any foreseeable incidents. In some cases, this requires that the confined space is cut open to facilitate rescue (e.g. a large metal sided silo might be cut open to create a large hole through which a casualty on a stretcher can be carried). Use of a harness with a rope securely attached to a retrieval line should be considered. The other end of the retrieval line must be attached to a mechanical device or a fixed point outside the permit space. A mechanical device should be available to retrieve someone from vertical-type permit spaces more than five feet (1.5m) deep.

Fire Prevention

If flammable/explosive atmospheres are possible within the confined space then thought must be given to fire safety. Equipment taken into the space must be appropriate for use in the types of explosive atmosphere anticipated. For example, electric light taken into the space must be rated as safe to use in an explosive gas or dust environment. The explosion protection of electrical equipment is dealt with in more detail in Learning Outcome 11.3.

Lighting

Since the confined space is unlikely to be well lit, it will be necessary for workers to take light sources with them. These may be battery-powered hand-held or head-mounted torches or electric lamps supplied by flex at 110V or mains supply. In all cases, the light must be suitable for the environment of use and must be carefully positioned so as to be safe and give good illumination.

Worker Suitability

All workers put into the confined space to do the work must be competent to undertake that type of work. They must be competent in confined space entry and in the use of the specific items of PPE (such as BA) that they have to wear. They must also be trained and familiar with the relevant emergency procedures, such as fire evacuation, gas detection response, etc.

Workers must also be suitable in terms of their physical fitness to undertake the work and for the confined space entry. Workers required to wear BA will usually need to undergo medical examination to certify them as fit to wear BA. Workers entering a constricted space through very narrow openings must be of a suitable size so that they will not get stuck and endanger their own and others' safe evacuation of the space.

Workers must also be psychologically suitable to undertake the work. Workers who suffer from claustrophobia would be unsuitable to put into a small confined space because they might panic and start to behave irrationally in response to their phobia. Even putting the face piece of a BA set over their head might prove too much for such a worker. (This topic is covered in more detail in Unit 2, Learning Outcome 10.1.)

Emergency Procedures

Arrangements for rescue in an emergency should be prepared. The types of emergency response required will vary depending on the nature of the confined space and the work activity carried out within it. In general:

- Personnel and equipment should be readily available for emergencies.
- Equipment should consist of:
  - BA sets.
  - Resuscitators.
  - Means of summoning help, e.g. two-way radio.
  - Life-lines.
  - Oxygen for resuscitation.
- Rescue teams should not enter a confined space without BA.
- Rescue teams should be thoroughly trained in:
  - Rescue techniques.
  - First aid.
  - Use of resuscitators.

MORE...

Information on confined spaces can be found in the following publications:

Safe Work in Confined Spaces – Confined Spaces Regulations 1997 – Approved Code of Practice and Guidance available from the UK HSE at:

www.hse.gov.uk/pubns/priced/l101.pdf

Chapter 58 of the ILO's *Encyclopaedia of cccupational health and safety*, available at:

https://iloencyclopaedia.org/part-viii-12633/safety-applications/item/979-confined-spaces

Training for Work in Confined Spaces

Effective training is necessary for:

- Supervisors.
- Anyone likely to work in confined spaces.
- Anyone likely to act as an attendant (top man).
- Members of rescue teams.

In practice, these functions may overlap, but **all** persons must be trained.

Training should include:

- Equipment:
  - Methods of use, limitations, interpretation of results, maintenance and calibration.
  - Knowledge of its construction and working.
  - Dealing with malfunctions and failures during use.
- Check procedures when donning apparatus.
- Emergency procedures.
- Artificial resuscitation.

Those likely to be involved in an emergency rescue operation should have appropriate training to cover:

- Likely causes of an emergency.
- Use, maintenance and testing of rescue equipment.
- Resuscitation procedures.
- Initiation of emergency response.

To be fully effective, all training should be supplemented with regular drills, rehearsals or exercises. Over-reliance on classroom training alone does not allow workers to practice the physical skills that they will need to put into operation during normal work and in the event of an emergency.

Maintenance of Equipment

Inspections of breathing apparatus, lifelines, signal equipment and other equipment should be carried out before use, routinely (e.g. monthly) and after use.

A competent person should carry out the checks and enter at least the following details in a register:

- The name of the occupier of the place of work.
- The address of the place of work.
- Particulars of the apparatus (distinguishing number/description).
- The date of the examination and by whom it was carried out.
- The condition of the apparatus, belt or rope, and particulars of any defects found during the examination.

MORE...

More information on confined spaces is available in L101 *Safe work in confined spaces, Confined Spaces Regulations 1997, Approved Code of Practice and guidance* (3rd edition, 2014) available at:

www.hse.gov.uk/pubns/books/l101.htm

and in INDG258 *Confined spaces, A brief guide to working safely* which can be found at:

www.hse.gov.uk/pubns/indg258.htm

STUDY QUESTIONS

1. Identify the specific risks relating to confined spaces.
2. Identify three ways in which the atmosphere within a confined space can be hazardous to workers.
3. When is it possible for a worker to enter or work in a confined space without breathing apparatus?
4. What equipment should be available in case of an emergency?

(Suggested Answers are at the end.)

Summary

Confined Spaces

We have:

- Examined the meaning of confined spaces with reference to the **Confined Spaces Regulations 1997**.
- Given examples of where confined space entry may occur in the workplace, such as:
  - Pits in garages.
  - Trunking ducts.
  - Watercourses, trenches, tanks, silos and sewers.
- Outlined the factors to be considered when assessing risks, including safe access arrangements, likely atmospheres to be encountered (including oxygen-enriched, oxygen depleted, toxic and flammable atmospheres), the task, materials and equipment, people at risk, reliability of safeguards (including PPE).
- Examined the factors to be considered in designing safe working practices including operating procedures and emergency policy and procedures and training for work in confined spaces.
- Explained the importance of providing and maintaining a safe atmosphere, including provision of breathing apparatus.
- Outlined the risk assessment method to be followed, considering the person at risk, task, material, equipment and the reliability of existing safeguards.
- Outlined the design of safe working practices, particularly the need for a permit to work in a confined space.
- Outlined the requirement for emergency procedures and the provision of training before any work in a confined space.

Learning Outcome 11.3

NEBOSH National Diploma for Occupational Health and Safety Management Professionals

ASSESSMENT CRITERIA

- Describe the mechanisms for fire and explosions, how building materials behave in a fire and methods that can be used for prevention and protection from fire and explosion.

LEARNING OBJECTIVES

Once you've studied this Learning Outcome, you should be able to:

- Outline the properties of flammable and explosive materials and the mechanisms by which they ignite.
- Outline the behaviour of structural materials, buildings and building contents in a fire.
- Outline the main principles and practices of prevention and protection against fire and explosion.

Contents

Flammable and Explosive Materials and the Mechanisms by which they Ignite

IN THIS SECTION...

- The three components of the fire triangle - fuel, oxygen and an ignition source - are all required for a fire to start.
- Fire is a combustion reaction - fuel is converted to combustion products (smoke, fumes, gases) in the presence of oxygen.
- The relevant properties of solids, liquids and gases have an influence on combustion.
- The following terminology is essential in understanding the properties of combustible and potentially explosive solids, liquids and gases: flash point, fire point, auto-ignition temperature, vapour density, limits of flammability, maximum explosion pressure and rate of pressure rise.
- Three types of explosions that involve the explosive combustion of flammable gas/vapour clouds are:
  - Unconfined Vapour Cloud Explosion (UVCE).
  - Boiling Liquid Expanding Vapour Explosion (BLEVE).
  - Confined Vapour Cloud Explosion (CVCE).
- Mechanisms of explosions and fire spread include:
  - How an explosion/fire occurs.
  - The stages of combustion can be divided into the following five stages: induction, ignition, growth, steady state and decay.
  - Mechanisms of confined and unconfined vapour cloud explosions and boiling liquid expanding vapour explosions.
- An explosion is a sudden and violent release of energy, causing a pressure blast wave. Atomisation/particle size and/or oxygen content have an effect on the likelihood and severity of fire/explosion.
- There are five principles of control for safe working with flammable substances: ventilation, ignition, containment, exchange and separation and the failure of these control measures coupled with the physico-chemical properties of flammable materials can bring about an explosion.
- Control measures for entering flammable atmospheres include purging, to keep flammable atmospheres below Lower Explosive Limits (LEL).
- The principles of selection of electrical equipment for use in flammable/explosive atmospheres.
- The prevention and mitigation of vapour phase explosions may be achieved through structural protection; plant design and process control; segregation and storage of materials; hazardous area zoning (**Dangerous Substances and Explosive Atmospheres Regulations 2002 (DSEAR)** Schedule 2), inerting and explosion relief; controlling the amount of material; prevention of release; segregation and storage of materials; hazardous area zoning; control of ignition sources; and sensing of vapour between Lower Explosive Limit (LEL) and Upper Explosive Limit (UEL) in order to detect the formation of an explosive atmosphere.
- Most organic solids, most metals, and some combustible inorganic salts can form explosive dust clouds. The dust pentagon: fuel, oxygen, an ignition source, confinement of the reaction and dispersion of the fuel are all required for a dust explosion to occur.
- The mechanisms of dust explosions involve combustible solid particle size, dispersal, explosive concentrations, ignition energy, temperature and humidity.

- Dust suspended in air in process equipment can allow dust explosion conditions to occur (a primary explosion). Secondary explosions are caused when lying dust in the workplace is disturbed and forms a second dust cloud which is then ignited by the heat released from the primary explosion.
- The prevention and mitigation of dust explosions include relevant hazardous places zoning (**DSEAR** Schedule 2).

The Fire Triangle

What do we need to start a fire?

1. A **combustible substance or fuel** (wood, paper, plastics, etc.).
2. **Oxygen** in a gas state (usually from air).
3. An **ignition source** (or heat).

If the conditions are right and these three factors are present, the substance will catch fire, i.e. heat and light will be evolved, accompanied by volumes of smoke and gases which will rise away from the fire.

The fire triangle

The three factors above form the basis of the 'fire triangle'; all must be present to produce and sustain a fire.

Fuel

Fuel consists of flammable and combustible materials which cover all states of matter. They include:

- Combustible solids, e.g. wood, plastics, paper and wrapping and packaging materials, soft furnishings and fabrics and even metals, e.g. magnesium.
- Flammable solids, e.g. magnesium (in finely divided form, such as powder).
- Flammable liquids, e.g. petroleum and its derivatives, paints, solvents, oils, etc.
- Flammable gases, e.g. hydrogen, Liquefied Petroleum Gas (LPG), methane, etc.

Oxygen

Although under certain unusual circumstances it is possible to produce combustion-like chemical reactions with materials such as chlorine or sulphur, it is safe to say that nearly all combustion requires the presence of oxygen. The higher the concentration of oxygen in an atmosphere, the more rapidly burning will proceed.

While the most common source of oxygen is from the air, in some workplaces there may also be additional sources, e.g. oxygen cylinders, or substances that are oxidising agents.

Heat

Heating a very small quantity of fuel and oxygen mixture (to a sufficient degree) is enough to start a fire. Fires are by definition exothermic (i.e. they release heat), and the very small fire started by a tiny heat source supplies to its surroundings more heat than it absorbs, enabling it to ignite more fuel and oxygen mixture, and so on, until very quickly there is enough heat available to propagate a large fire. The heat may be provided by various sources of ignition.

Ignition Sources

Ignition sources include:

- Open flames - matches, welding torches, etc.
- Electrical sparking sources.
- Spontaneous ignition.
- Sparks from grinding or tools.
- Static electricity.
- Friction.
- Hot surfaces (e.g. heaters or overheating equipment).
- Sparks (arising from impact of metal tools, electrical arcing and static discharges).
- Lasers and other intense radiant heat sources.
- Chemical reactions giving rise to heat/flame.
- Smoking.

Potential ignition source

Properties of Solids, Liquids and Gases

In order to be able to assess the risk of fire and explosion, we need to understand how fires and explosions occur. This requires some knowledge of the properties of combustible and potentially explosive solids, liquids and gases and in particular, the following important terms.

Simply defined, fire is a chemical reaction in a mixture of gases. For the most part, fire is a mixture of hot gases and the flames are the result of a chemical reaction, primarily between oxygen in the air and a fuel, such as wood, a flammable liquid or propane gas. In addition to other products, the reaction produces carbon dioxide, steam, light, and heat.

Fires start when a flammable or a combustible material, in combination with a sufficient quantity of an oxidiser such as oxygen gas or another oxygen-rich compound, is exposed to a source of heat or ambient temperature above the flash point for the fuel/oxidiser mix, and is able to sustain a rate of rapid oxidation that produces a chain reaction.

This is commonly called the 'fire tetrahedron'. Fire cannot exist without all of these elements in place and in the right proportions. For example, a flammable liquid will start burning only if the fuel and oxygen are in the right proportions. Some fuel-oxygen mixes may require a catalyst, a substance that is not consumed, when added, in any chemical reaction during combustion, but which enables the reactants to combust more readily.

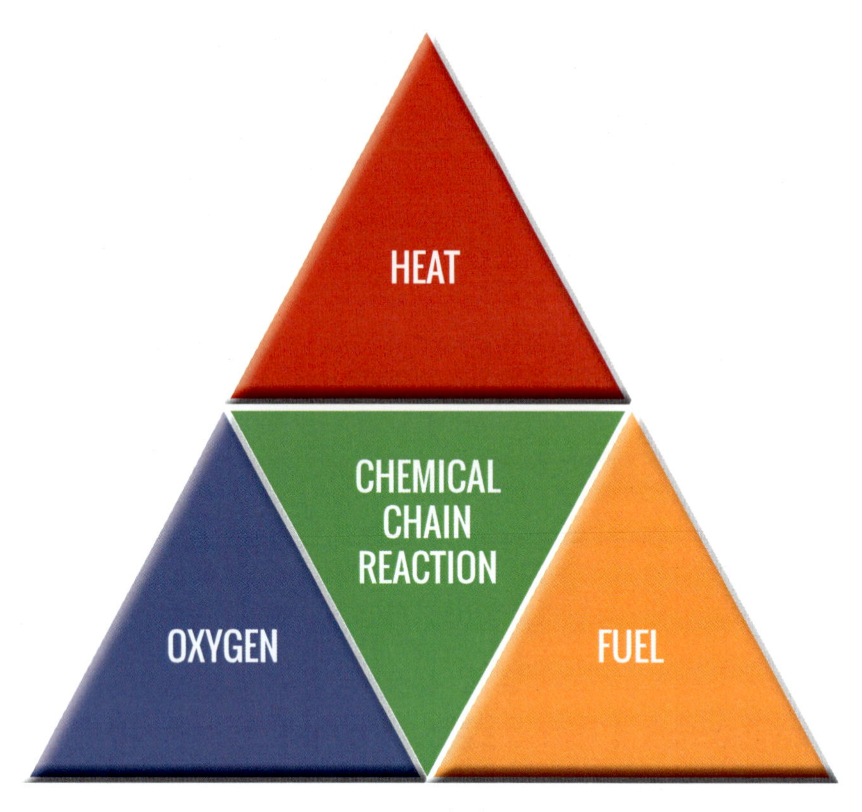

Fire tetrahedron

Once ignited, a chain reaction must take place whereby fires can sustain their own heat by the further release of heat energy in the process of combustion and may propagate, provided there is a continuous supply of an oxidiser and fuel.

Flash Point

DEFINITION

FLASH POINT

The lowest temperature of a liquid at which sufficient vapour is given off to ignite momentarily (flash), when an external source of ignition is applied.

Note that sustained combustion does not occur.

Substances which have a flash point below ambient temperature will pose a significant fire risk as they will be producing a flammable vapour at ambient temperatures.

Some common solvents and their flash points

Solvent	Flash Point (°C) (Abel closed cup)
Butanone (methyl ethyl ketone, MEK)	−7
Carbon disulphide (CS_2)	−30 (Auto-ignition temp. 102°C)
Diesel oil	+40 (approx.)
Ethyl ethanoate (ethyl acetate)	−4
Ethoxyethane (diethyl ether)	−40
Methylated spirit	10
Methylbenzene (toluene)	4
Petrol	−40 (approx.)
Phenylethene (styrene)	32
Propanone (acetone)	−17

Note: Petrol is far more dangerous than diesel oil if it is spilled because the vapours from petrol will ignite from any fortuitous ignition source, whereas those from diesel will not. Contemplate such a spillage into the bilges of a boat and you can see why the insurance premiums are higher if the boat has a petrol engine.

Substances with very low flash points are very volatile. Ether, acetone and carbon disulphide are all notoriously dangerous.

Flash point tests may be applied to **any** liquid and not just hydrocarbons. There are several types of apparatus which can be used for the determination of flash point, such as the Abel Closed Cup Apparatus.

Fire Point

DEFINITION

FIRE POINT

The lowest temperature of a liquid at which sufficient vapour is given off at the surface that the application of an external ignition source will lead to continuing burning.

Note that sustained combustion does occur.

The fire point temperature is usually just above the flash point. Often 10°C or so above.

Note that technically both flash point and fire point are temperatures of materials in the **liquid** state. The **flash point** of a liquid is the temperature of greatest interest since that is the temperature that is then used to classify the liquid as flammable. So, for example, in the EU classification of flammable liquids the category 1 (extremely flammable), 2 (highly flammable) or 3 (flammable) is determined using the flash point of the liquid (thus petrol is a highly flammable liquid, but diesel is not).

Auto-Ignition Temperature (AIT)

DEFINITION

AUTO-IGNITION TEMPERATURE

The lowest temperature at which a substance will ignite without the application of an external ignition source.

Once the Auto-Ignition Temperature (AIT) has been reached, there is no requirement for an external ignition source to achieve combustion. The material will 'spontaneously' ignite. As an example, methane gas has an AIT of 580°C in air. Thus, a mixture of methane gas and air at 580°C will burn simply because it is at that temperature. It auto-ignites.

The fuel can be a solid, liquid or gas and once ignition has taken place, the material will sustain the self-ignition in the absence of spark or flame.

The value is influenced by the material's size and the shape of the heated surface and, in the case of a solid, the rate of heating and other factors.

A chemical or biochemical reaction can supply the heat to raise a substance above its AIT. Reactive chemicals can auto-ignite when mixed together due to the exothermic (heat-generating) nature of the chemical reaction. Haystacks have been known to auto-ignite due to bacteriological action causing internal heating.

On a more practical note, the relatively low AIT of diesel, which is, surprisingly, lower than petrol, means that diesel engines do not have to have a spark plug. The action of compressing the fuel/air mixture in the cylinders of the engine is enough to raise it above the AIT and cause ignition.

Vapour Density

DEFINITION

VAPOUR DENSITY

The mass of vapour per unit volume (its weight).

Vapour is the dispersion of molecules of a substance in air that is liquid or solid in its normal state, e.g. water vapour, benzene vapour.

('Normal state' means at normal temperature and pressure.)

Vapour density gives a measure of how much vapour there is in the air.

Limits of Flammability

These limits of flammability apply specifically to gases and vapour mixed with air, such as methane gas mixed with air and acetone vapour mixed with air.

DEFINITIONS

LIMITS OF FLAMMABILITY

The extremes of fuel (vapour or gas) to air ratios between which the mixture is combustible.

LOWER FLAMMABLE LIMIT (LFL) also known as the **LOWER EXPLOSIVE LIMIT (LEL)**

The lowest percentage of gas/vapour to air mixture that will sustain combustion.

UPPER FLAMMABLE LIMIT (UFL) also known as the **UPPER EXPLOSIVE LIMIT (UEL)**

The highest percentage of gas/vapour to air mixture that will sustain combustion.

FLAMMABLE RANGE

The spread of percentage between a mixture's LFL (LEL) and UFL (UEL).

STOICHIOMETRIC MIX (RATIO OR POINT)

The gas/vapour to air mixture that gives the most complete and efficient combustion.

Below the LFL, the mixture has too much air and not enough fuel (gas or vapour) to allow combustion to take place (lean or weak mixture). Above the UFL, the mixture has too much fuel in proportion to the air for combustion to take place (rich mixture).

These figures are important in helping to decide suitable fire prevention strategies; it is usual to try to keep the atmosphere below the LFL (LEL).

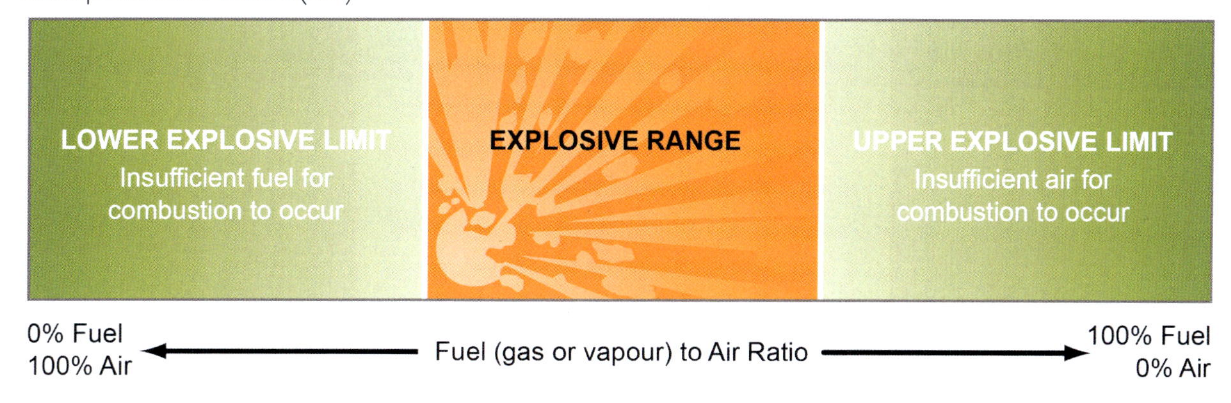

Flammable limits

The following table shows the limits for some flammable liquids and gases in air:

Examples of flammable limits in air

Fuel	Physical State	Lower % Limit	Upper % Limit
Hydrogen	Gas	4	74
Carbon monoxide		12.5	74
Methane		5	15
Propane		2.5	9.5
Butane		2	8.5
Ethylene (ethene)		3	32
Acetylene (ethyne)		2.5	80
Ethyl alcohol (ethanol)	Vapour from liquid	4	19
Carbon disulphide		1.3	50
Petrol		1.4	7.5
Paraffin		0.7	5.0
Diethyl ether (ethoxyethane)		2	36

In pure oxygen or oxygen-enriched atmospheres, the limits become wider than they are for air - so it is more likely for a mixture of gas or vapour to be within the flammable range, and fires are correspondingly more difficult to extinguish.

Conversely, if the air or oxygen is diluted with an inert gas such as nitrogen or carbon dioxide, the limits become narrower until they converge together and there is no flammable range. Such dilution would typically reduce the oxygen content to 8%-10% (the same limit below which an atmosphere would be irrespirable and unable to support life).

So just to make the point - if you want to inert the vapour space at the top of a storage tank full of methanol liquid (a highly flammable liquid with a flash point of 12°C), you do not need to pump in nitrogen to achieve 0% oxygen in that vapour space. You simply need to pump in sufficient nitrogen gas to dilute the oxygen concentration to below 10%. You can then flick matches in that tank all day long and no ignition will take place (although you would need BA to survive in there).

In the practical situation of a leak involving petrol, dispersion and natural dilution in still air would reduce the vapour content below the LFL at a distance of about 12m from the leak; this is known as the **safe dilution point** and gives an indication of the size of zone which would be needed to control sources of ignition in order to prevent a fire.

UVCEs, CVCEs and BLEVEs: Causes and Effects

We are now going to look at three different types of explosions that involve the explosive combustion of flammable gas/vapour clouds. They are called Unconfined Vapour Cloud Explosions (UVCEs), Confined Vapour Cloud Explosions (CVCEs) and Boiling Liquid Expanding Vapour Cloud Explosions (BLEVEs).

Unconfined Vapour Cloud Explosion (UVCE)

A UVCE results from the release of a considerable quantity of flammable gas or vapour into the atmosphere, and its subsequent ignition. Such an explosion can cause extensive damage, such as occurred at Flixborough in 1974.

MORE...

A summary of the Flixborough accident can be found here:

www.hse.gov.uk/comah/sragtech/caseflixboroug74.htm

The UCVE at Buncefield storage depot in 2005 is an excellent case study. The HSE website has a lot of information relating to this event and the trial that followed:

www.hse.gov.uk/comah/buncefield

Boiling Liquid Expanding Vapour Explosion (BLEVE)

A BLEVE involves a sudden release of vapour due to the failure of a pressure vessel. A BLEVE can occur with both flammable and non-flammable liquids (e.g. water in a boiler). However, the most common and damaging BLEVEs involve LPG storage vessels, such as propane gas storage tanks and road or rail tankers/cars:

- Sections of the metal vessel are thrown out in the explosion.
- A blast wave is created causing massive structural damage.
- The entire remaining contents of the tank are released into the atmosphere resulting in the sudden release of super-heated LPG to vapour that then ignites causing a huge fireball to rise up on a thermal current. This causes heat radiation damage over significant distances and starts secondary fires.

For example, a BLEVE of a propane sphere of 15m in diameter could cause damage as far away as 4,500m, and radiation and fragmentation damage would each extend to about 1,000m.

MORE...

Videos of LPG tank BLEVEs are easy to find on the internet.

One incident of note is the multiple-BLEVE incident that occurred outside Mexico city in 1984 involving spherical LPG tanks that lead to the deaths of 500-600 people and thousands of severe burn injuries.

Confined Vapour Cloud Explosion (CVCE)

If a flammable vapour cloud is ignited in a container, e.g. a process vessel, or in a building - so that it is confined - pressure builds up until the containing walls rupture. This is a CVCE. As with natural gas explosions in buildings, a relatively small amount of flammable material (a few kilograms) can lead to an explosion when released into the confined space of a building.

CVCEs can cause considerable damage, e.g. peak over-pressures of up to eight bars can be experienced in a fully confined explosion, and much higher in the unlikely event of a detonation, but in general they have insufficient energy to produce more than localised effects as far as off-site damage is concerned (e.g. broken windows). Much of the major damage found within the confines of the cloud following a UVCE probably results from local CVCEs. If the results of a CVCE affect nearby plant or equipment, serious secondary explosions can follow.

For personnel close to the blast, the pressure wave itself, missiles and flash-burns can result in serious or fatal injuries. For instance, fatalities have occurred from explosions during hot work on inadequately cleaned/purged 45-gallon drums which had contained flammable residues, or in natural gas or LPG explosions following leaks into rooms.

Mechanisms of Explosions and Fire Spread

Explosions

DEFINITION

EXPLOSION

A sudden and violent release of energy, causing a pressure blast wave.

Usually it is the result, not the cause, of a sudden release of gas under high pressure, but the presence of a gas is not necessary for an explosion.

An explosion may occur from a physical or mechanical change or from a chemical reaction. An explosion can occur without combustion, such as the failure through overpressure of a steam boiler or an air receiver.

In discussing explosions involving combustion we must distinguish between a **detonation** and a **deflagration**:

- If a mixture **detonates**, the reaction zone propagates at **supersonic velocity** and the principal heating mechanism of the mixture is **shock compression**.
- In a **deflagration**, the combustion process is the same as in the normal burning of a gas mixture; the combustion zone propagates at **subsonic** velocity and the **pressure build-up is slow**.

When a material (such as a high explosive) **detonates**, a shock wave moves through the material and it is this pressure wave (causing compression and heating) that causes the further combustion reaction to take place. In effect, the shock wave produced by the exploding material causes the material to react more; which causes more of a shock wave; which in turn causes more reaction to occur. So a positive feedback loop is created. The key thing is - the shock wave causes the reaction.

An escaped cloud of petrol vapour is far more likely to **deflagrate**. The vapour cloud comes into contact with an ignition source and then combustion of the vapour takes place at subsonic speeds. The shock wave travels out at the speed of sound, but the combustion reaction zone moves more slowly. In effect, the flames travel more slowly through the expanding cloud of petrol vapour than the blast wave does.

Whether detonation or deflagration occurs in a gas/vapour-air mixture depends on various factors, including the concentration of the mixture and the source of ignition. Unless confined or ignited by a high-intensity source (a detonator), most materials will not detonate. However, the pressure wave (blast wave) caused by a deflagration can still cause considerable damage. The **Buncefield** petrol vapour explosion that occurred in Hertfordshire in 2005 was a deflagration. The blast wave was so big that the boom was heard in the Netherlands.

Certain materials, such as acetylene, can decompose explosively in the absence of oxygen and because of this are particularly hazardous.

Maximum Explosion Pressure

DEFINITION

MAXIMUM EXPLOSION PRESSURE

The maximum pressure occurring in a closed vessel during the explosion of an explosive atmosphere. It is determined under specified test conditions with different atmospheres (dust, gas, etc.) having different maximum explosion pressure values.

If an explosion takes place in an unvented vessel, the rise in pressure to the maximum explosion pressure may cause damage to the vessel and surrounding area. To produce a vessel that would withstand the maximum pressure of an explosion would probably result in an uneconomic, heavy, thick-walled structure.

The alternative is to provide venting in the form of bursting discs or panels, at a pressure below the maximum explosion pressure. On reaching this pressure, the vents will open and dissipate the pressure before excessive damage is caused. (This issue will be discussed later in this Learning Outcome.)

Rate of Pressure Rise

DEFINITION

RATE OF PRESSURE RISE

The rate of the pressure rise in a given time during the explosion of an explosive atmosphere in a closed vessel. This is determined under test conditions.

The damage caused as a result of the blast pressure front of an explosion is a result of two factors: the maximum explosive pressure and the rate of pressure rise. If the pressure rise is rapid, then the containing vessel will not have sufficient time to resist the forces. The need for venting is therefore important. The data from specified tests to determine maximum explosion pressure and rate of pressure rise are used to design explosion mitigation methods considered later in this Learning Outcome.

DEFINITION

FIRE

A combustion reaction in which fuel is converted to combustion products (smoke, fumes, gases) in the presence of oxygen. It is a rapid, self-sustaining, gas-phase oxidation process which produces heat and light.

When combustion takes place 'in' solids or liquids, it is the vapours given off which ignite rather than the solid or liquid itself. So the combustion reaction actually takes place at or above the surface of the solid or liquid.

In basic terms, when a fire is burning (combustion is taking place) volatile molecules of the fuel are combined with oxygen to produce new compounds (combustion products). This is an **oxidation** reaction.

Fuel does not generally spontaneously combust in air (think of a lump of coal). It requires some energy to vaporise sufficient fuel molecules to initiate the reaction, e.g. by supplying heat. Once initiated, the heat produced by the reaction itself can supply the heat needed to sustain further vaporisation, ignition and combustion of fuel, so that the external heat source is no longer required.

The Stages of Combustion

Combustion can be divided into the following five stages:

- **Induction**

  Heat is initially supplied by an external source which results in heating of the fuel and the production of flammable vapours. These vapours mix with air above the fuel.

- **Ignition**

  The point of ignition is reached when the flammable vapours are hot enough to combust; the vapours given off by the fuel are now chemically reacting with the oxygen (i.e. an oxidation reaction is taking place) in the air to produce heat and combustion products. Thus the reaction becomes self-sustaining (and no longer requires an external heat source).

- **Growth**

  At this stage combustion develops very quickly and there is a dramatic increase in temperature as the fire grows. The fire may spread either through direct burning or through the typical mechanisms of heat transmission (convection, conduction or radiation). The rate, scale and pattern of growth depend on a number of factors such as the:

  - Nature, form and amount of the fuel.
  - Availability of oxygen (open, ventilated versus sealed containment).
  - Amount of heat produced by the reaction.

- **Steady State**

  After the growth period the temperature stabilises and the combustion process reaches a steady state where the reaction between fuel and oxygen is balanced until all the fuel is consumed. The fire may still be physically spreading, but the rate of burning and temperatures reached are not increasing.

- **Decay**

  Decay will begin when either the fuel or oxygen has been consumed. The rate at which the fire is burning and the maximum temperatures decline. The fire will eventually extinguish and then gradually cool. In the early stages of decay, there is still a considerable amount of heat; there is certainly enough to cause re-ignition if more fuel or oxygen is supplied. In the latter case, admission of oxygen (e.g. opening a window) into an oxygen-depleted room can result in the sudden explosive re-ignition of vapours.

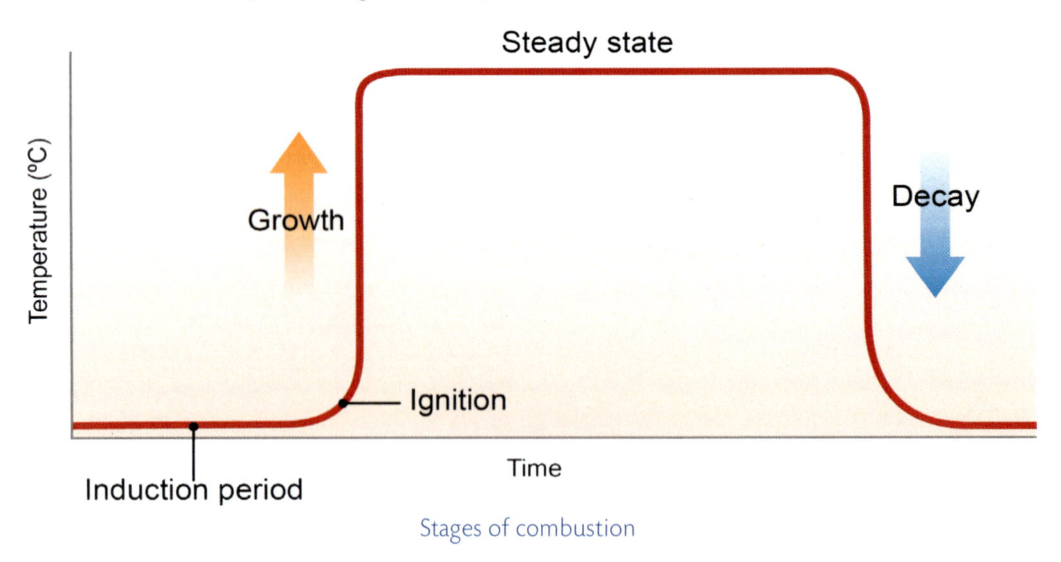

Stages of combustion

Mechanisms of UVCEs, CVCEs and BLEVEs

Unconfined Vapour Cloud Explosions (UVCEs)

When a large amount of volatile material is released rapidly into the atmosphere, a vapour cloud forms and disperses. If the cloud is ignited before it is diluted below its LFL, an unconfined vapour cloud explosion will occur. This is one of the most serious hazards in the process industries. Both shock waves and thermal radiation will result from the explosion; the shock waves will usually produce the greater damage.

The energy of the blast wave is generally only a small fraction of the energy available from the combustion of all the material that constitutes the cloud. The ratio of the actual energy released to that available is often called the **explosion efficiency**.

Unconfined vapour clouds can deflagrate or detonate, but a deflagration is much more likely. A detonation is more destructive, but a deflagration can produce a damaging pressure wave.

UVCEs require a release of flammable liquid, vapour or gas, moderate dispersion to produce a very large flammable vapour/air cloud, and an ignition source.

Confined Vapour Cloud Explosions (CVCEs)

CVCEs occur when a flammable vapour cloud (within its flammable or explosive range), ignites in a closed space such as a process vessel, pipe or building. The resulting expansion of the gases involved in the chemical reaction can result in pressure building up inside until the containing walls rupture.

A relatively small amount of flammable material can often lead to a significant explosion because of the restriction imposed on the reaction by the walls of the vessel or building, and can cause considerable damage. This has been illustrated in the case of both the fatal CVCE at the ICL Plastics in Glasgow.

CASE STUDY

The CVCE that caused the structural collapse of an ICL Plastics Ltd mill building in Glasgow in 2004, killing nine people and injuring 33, is a useful case study. The explosion followed the release of a small quantity of propane gas into the basement of the building as a result of a gas pipe fracture. ICL Plastics and ICL Tech were subsequently fined a total of £400,000 over the incident. A report into the explosion was published following a public inquiry. This report is easy to find by searching "ICL report" on the HSE website.

Boiling Liquid Expanding Vapour Explosions (BLEVEs)

LPG storage tanks and tankers of any significant size are pressure-relieved. This means the tank is fitted with a pressure-relief valve so that if the tank gets hot, it will not rupture due to overpressure (in the first instance) but instead the pressure relief valve will open, venting off excess pressure from the tank.

In order for a relieved tank of this type to explode it needs to be subjected to extreme heat - usually in a concentrated area of the vessel shell. This might occur if a building or vehicle fire were to occur adjacent to the storage tank. The flames of the fire heat the LPG tank - the metal walls of the tank heat up, but in the first instance, the LPG liquid inside the tank acts as a heat sink. So the metal cannot get so hot that it starts to lose its strength (in much the same way that if you boil a saucepan full of water the walls of the pan do not get hotter than 100°C because that is the boiling point of the water). Instead, the LPG inside the tank heats up, vapourises and the pressure inside the tank increases. This activates the pressure relief valve and LPG is vented straight to atmosphere. This vented LPG would normally catch fire, but as it is vented straight up into the air it is unlikely to cause any damage.

However, as the tank vents its contents, the level of liquid in the tank decreases. Eventually, the tank will have little liquid in it, which means the flames from the fire are now in contact with the metal sides of the tank and there is no liquid heat sink on the other side of that metal (the saucepan has been boiled dry). The metal now gets so hot that it starts to become structurally weakened and the pressure inside the tank causes the metal to catastrophically fail:

- Sections of the metal vessel are thrown out in the explosion.
- A blast wave is created causing massive structural damage.
- The entire remaining contents of the tank are released into the atmosphere resulting in the sudden release of super-heated LPG to vapour that then ignites causing a huge fireball to rise up on a thermal current. This causes heat radiation damage over significant distances and starts secondary fires.

For example, a BLEVE of a propane sphere of 15m in diameter could cause damage as far away as 4,500m, and radiation and fragmentation damage would each extend to about 1,000m.

Effects of Atomisation/Particle Size

Atomisation (when referred to liquid) is a process where the liquid is turned into tiny liquid particles or droplets (typically by use of an aerosol spray or injector). This has the effect of allowing far better mixing with air as each droplet is surrounded by air (as opposed to a container of liquid where only the surface comes into contact with the air). In addition, the tiny droplets have a much greater surface area that enables reaction with the oxygen in the air. Consequently, should a source of ignition be applied to an atomised fuel/air mixture, the resulting combustion is far more rapid as each droplet will ignite almost instantaneously.

A good example of this is when the walls of an LPG vessel fail during a BLEVE. The liquid in the vessel has been absorbing the heat from the fire and any pressure build-up inside the vessel has been relieved by the pressure relief valve but when the vessel fails, any remaining liquid is immediately exposed to ambient pressure and temperature. The LPG, which remained in a liquid state because it was under pressure, now expands exponentially (LPG expansion is 270 times the volume of gas to the volume of liquid) and becomes a cloud of LPG droplets (the fuel is 'atomised') which are ignited by the source of the fire. You will notice that in any film of a BLEVE that the 'flame front' seems to move relatively slowly through the cloud of droplets, which is why this is often known as a 'deflagration - a subsonic explosion'. The flame is, literally, jumping from one droplet to the next - obviously at great speed!

An effect of this is that very high radiated temperatures are generated close to the burning cloud capable of incinerating any organic objects and damaging plant and equipment in the vicinity.

The principle of fuel atomisation is used in car engines to ensure rapid burning of the fuel/air mixture.

When it comes to solids, the equivalent of atomisation (for liquids) is breaking them down into small particles (like dusts). Breaking things down into smaller particles gives a greater surface area and makes it easier to burn when mixed with air.

This is why it is not possible to light a log of wood in a cold log burner with a match. But it is possible to light paper that can then be used to light kindling (small shavings and twigs of wood) that can then be used to light larger sticks that can in turn be used to set light to the log.

How Failure of Control Measures can Bring About an Explosion

An explosion can occur when:

- Flammable materials produce a concentration of gas or vapour in air which is above the LFL (LEL) and below the UFL (UEL), i.e. in the flammable or explosive range.
- A source of ignition of sufficient energy is present to ignite the vapour.

The UK HSE publication INDG227 *Safe working with flammable substances*, (no longer available on the HSE website, but still relevant) establishes five principles of control for safe working with flammable substances which aim to prevent fires and explosions:

- **Ventilation** ensures that any vapours given off from a spill, leak or release from any process will be rapidly dispersed, preventing the formation of a vapour/air mixture above the LFL.
- **Ignition** sources must be removed from storage and handling areas for flammable materials. If the ignition source generates energy above the **Minimum Ignition Energy (MIE)** for the combustible vapour or gas then an explosion will occur.
- **Containment** prevents the escape of flammable materials into the workplace which can release vapours and generate explosive mixtures.
- **Exchange** of a flammable substance for a less flammable one will reduce the risk of formation of an explosive mixture. Flammable liquids with flash points well above room temperature will be unlikely to form explosive atmospheres under normal working conditions.
- **Separation** of flammable substances from other processes and general storage areas by physical barriers, walls or partitions will contribute to a safer workplace by controlling the zone in which flammable atmospheres may be present.

These form a nice mnemonic: **VICES**.

Failure of these control measures, coupled with the physico-chemical properties of flammable materials, can bring about an explosion.

Process of Oxidisation

Oxidisation is the process of oxidising; the addition of oxygen to a compound. Combustion is a rapid, self-sustaining, gas-phase **oxidation** process which produces heat and light. Volatile molecules of the fuel are combined with oxygen to produce new compounds (combustion products). The potential combustion products depend on the fuel (and its complexity) as well as the completeness of the combustion reaction. If we take a simple case such as the combustion of carbon (e.g. charcoal), combustion products would be carbon dioxide (complete combustion) and carbon monoxide (incomplete combustion). Incomplete combustion occurs where the oxygen is limited.

TOPIC FOCUS

Chemical Equations

Chemical reactions are depicted as **chemical equations** to explain what is happening.

In a **chemical equation** the tiny entities (**molecules**) that react with each other are shown as **chemical formulae**, e.g. $\mathbf{H_2O}$ represents water, $\mathbf{CO_2}$ represents carbon dioxide.

Sometimes more than one **molecule** of a substance reacts with another one in a reaction so this is shown by a number in front of the **formulae**, e.g. **2**H_2O represents two molecules of water, CO_2 represents one molecule of carbon dioxide (we don't need to show the **1**).

When methane gas (CH_4) burns in air (i.e. reacts with oxygen, O_2) we find that **one molecule** of methane (CH_4) reacts with **two molecules** of oxygen ($2O_2$).

This reaction produces **one molecule** of carbon dioxide (CO_2) and **two molecules** of water ($2H_2O$).

This can be represented by the **chemical equation**:

$$CH_4 + 2O_2 \rightarrow CO_2 + 2H_2O$$

which is a simple form of shorthand to show what **molecules** are reacting together and **how many** of them are involved.

Complete Combustion

In a complete combustion reaction, fuel reacts with oxygen and the products are compounds of each element in the fuel with oxygen. For example, methane burns in air to form carbon dioxide and water:

$$CH_4 + 2O_2 \rightarrow CO_2 + 2H_2O$$

Another example is the combustion of hydrogen and oxygen to form water vapour:

$$2H_2 + O_2 \rightarrow 2H_2O + \text{heat}$$

Incomplete Combustion

Incomplete combustion occurs when there isn't enough oxygen to allow the fuel to react completely with the oxygen. When methane burns in limited air, the reaction will yield carbon dioxide and water, but also carbon monoxide and pure carbon (soot or ash). This results from the reduced oxygen supply and the inability for full oxidation to carbon dioxide and water:

$$4CH_4 + 6O_2 \rightarrow CO_2 + 2CO + C + 8H_2O$$

The Effects of Oxidising Substances on Fire and Explosion Mechanisms

Some chemicals have a relatively large amount of oxygen chemically bound within their molecular structure. Examples include:

Oxidising symbol

- Ammonium Nitrate; NH_4NO_3 (a common fertiliser).
- Sodium Chlorate; $NaClO_3$ (a banned weed killer).
- Hydrogen Peroxide; H_2O_2 (a bleaching agent).

Because these chemicals contain a large amount of oxygen, they are classified as **oxidising substances** or **strong oxidisers**.

If these substances are present during a fire they will release the oxygen that they contain. This adds more oxygen to the combustion oxidation reaction, which increases the growth rate of the fire. The end result is that combustion takes place more rapidly and higher temperatures are achieved. If the right conditions exist, then the rate of combustion may reach explosive proportions.

Strong oxidisers are often incompatible with other chemicals and, if mixed, can cause a violent exothermic reaction. In these cases there is no need for an external heat source to be applied to start the fire. You simply mix the chemicals together and the chemical reaction takes over. Again, if sufficient quantities of the right chemicals are mixed together this reaction would happen explosively.

For this reason, the storage of oxidising chemicals must be carefully managed to ensure that such chemicals are not exposed to significant heat sources and to ensure that they are not stored adjacent to incompatible chemicals. The fire protection of storage areas is a key concern.

Safe storage of oxidising materials and other chemically reactive materials is dealt with in more detail later in the unit.

Flammable Atmospheres

Flammable atmospheres can arise or be caused by the release of flammable gases, mists, vapours or combustible dusts into an area where, if there is enough of the substance, mixed with air and it is within the flammable or explosive range for that material, then all that is required is a source of ignition (at its Minimum Ignition Energy) to cause a fire or an explosion.

Many workplaces may contain potentially explosive atmospheres or have activities that produce such atmospheres. Examples include:

- Activities that create or release flammable gases or vapours, such as vehicle solvent paint spraying.
- Workplaces handling fine organic dusts such as grain flour - in bakeries or flour or sugar mills
- Woodworking shops or wood chipping facilities.
- Manufacturing facilities producing metal powder.
- Rubber tyre factories and tyre recycling facilities.
- Biomass energy facilities - anaerobic digestion plant (methane gas).
- Water pumping stations.

MORE...

A summary of the Abbeystead water pumping station explosion can be found here:

www.hse.gov.uk/comah/sragtech/caseabbeystead84.htm

Control Measures for Entering Flammable Atmospheres

Control measures for entering areas where a flammable atmosphere may exist include:

- carrying out a thorough risk assessment;
- employing safe systems of work;
- use of a permit to work;
- a competent and well trained workforce; and
- cleaning and purging of the area to keep any flammable atmosphere to below the LFL/LEL.

Ignition sources should be controlled and the workforce should use anti-static clothing, spark-proof (bronze) tools and intrinsically safe, explosion-proof (ATEX) electrical equipment.

DEFINITION

ATEX

The name commonly given to the two European Directives for controlling explosive atmospheres:

- Directive 99/92/EC (also known as 'ATEX 137' or the 'ATEX Workplace Directive') legislates on minimum requirements for improving the health and safety protection of workers potentially at risk from explosive atmospheres. It has been transposed into UK law as the **Dangerous Substances and Explosive Atmospheres Regulations 2002 (DSEAR)**.
- Directive 2014/34/EU (also known as the 'ATEX Equipment Directive') concerns equipment and protective systems intended for use in potentially explosive atmospheres. It has been transposed into UK law as the **Equipment and Protective Systems Intended for Use in Potentially Explosive Atmospheres Regulations 2016**.

Selection of Electrical Equipment for Use in Flammable Atmospheres

In situations where flammable atmospheres are present or likely to occur occasionally, particular precautions are necessary to prevent electrically-caused ignition of the atmosphere. Hot surfaces, arcs and sparks associated with electrical equipment can ignite such atmospheres, causing fires and explosions. To prevent such ignition, fire and explosion:

- Areas where such flammable atmospheres might exist must be recognised.
- Electrical equipment for use in such areas must be selected to ensure that it is suitable and will not cause fire or explosion.

As we know from earlier sections, flammable atmospheres are created by the presence of flammable gases, vapours and dusts. For an atmosphere to be capable of ignition, the concentration of the flammable substance in air must be at a certain level. The ignitable concentration lies between the upper and lower explosive limits (UEL and LEL) of the substance. Mixtures outside this range, i.e. rich and lean mixtures, will not ignite.

Categories of Electrical Equipment

Electrical equipment for use in ATEX-zoned atmospheres is classified by the manufacturer into one of three categories: category 1, 2 or 3:

- **Category 1** equipment is suitable for use in a Zone 0, 1 or 2 area or a Zone 20, 21 or 22 area.
- **Category 2** equipment is suitable for use in a Zone 1 or 2 area or a Zone 21 or 22 area.
- **Category 3** equipment is suitable for use in a Zone 2 area or a Zone 22 area.

Thus, Category 1 equipment gives the highest level of protection and category 3 the lowest. These standards are set out in the **Equipment and Protective Systems Intended for Use in Potentially Explosive Atmospheres Regulations 2016**. Schedule 3 of the **Dangerous Substances and Explosive Atmospheres Regulations 2002 (DSEAR)** states that equipment and protective systems must be selected on the basis of these regulations.

The methods used by the manufacturer of the electrical equipment to make it safe to use in the zoned environment vary but fall into a number of different techniques, each of which is indicated by a 'type of protection' code on the electrical equipment itself:

Type of protection	Basic principle
Flameproof enclosure (d)	Parts which can ignite a potentially explosive atmosphere are surrounded by an enclosure which can withstand the pressure of an explosive mixture exploding inside of it and prevents the propagation of the explosion to the atmosphere surrounding the enclosure. It can be used in Zones 1 or 2.
Increased safety (e)	Additional measures are taken to increase the level of safety, thus preventing the possibility of unacceptably high temperatures and the creation of sparks or electric arcs within the enclosure or on exposed parts of electrical apparatus, where such ignition sources would not occur under normal operation.
Pressurised apparatus (p)	The formation of a potentially explosive atmosphere inside a casing is prevented by maintaining a positive internal pressure of inert gas in relation to the surrounding atmosphere and, where necessary, by supplying the inside of the casing with a constant flow of inert gas which acts to dilute any combustible mixtures.*
Intrinsic safety (i)	Apparatus used in a potentially explosive area contain intrinsically safe electric circuits only. An electric circuit is intrinsically safe if no sparks or thermal effects are produced under specified test conditions (which include normal operation and specific fault conditions) which might result in the ignition of a specified potentially explosive atmosphere.
Oil immersion (o)	Electrical apparatus (or parts of it) are immersed in a protective fluid (such as oil), such that a potentially explosive atmosphere existing over the surface or outside of the apparatus cannot be ignited.
Powder filling (q)	Filling the casing of an electrical apparatus with a fine granular packing material, which makes it impossible for an electric arc created within the casing under certain operating conditions to ignite a potentially explosive atmosphere outside of it. Ignition must not result from either flames or raised temperature on the surface of the casing.
Encapsulation (m)	Parts which may ignite a potentially explosive atmosphere are embedded in a sealing compound, such that the potentially explosive atmosphere cannot be ignited.
Type of protection (N)	Electrical apparatus is not capable of igniting a potentially explosive atmosphere (under normal operation and under defined abnormal operating conditions).

* Purging is the process of flushing hazardous gas or dust out of an enclosure prior to the introduction of the inert gas. The process of making an electrical enclosure safe for use in an explosive atmosphere is therefore often referred to as purging and pressurising.

The type of protection appears on the electrical equipment alongside an EEX precursor, an Ex symbol and the category of the equipment.

Thus, Category 2 equipment of Type ib (see below) is suitable for use in a Zone 1 or 2 area (or 21 or 22) and is intrinsically safe.

'Intrinsically safe' means that the energy level of the equipment is insufficient to produce an incendiary spark. Two categories of intrinsically safe equipment exist:

Ex symbol

- 'ia' which is more stringent as it allows for two simultaneous faults; and
- 'ib' which allows for only one.

Only 'ia' equipment can be used (exceptionally) in Zone 0 if sparking contacts are not part of the equipment. Examples of type 'i' are instrumentation and low-energy equipment.

Control Measures to Prevent and Mitigate Vapour Phase Explosions

Investigation and analysis of the cause and consequences of vapour phase explosions identifies the following important control principles:

- **Structural protection** incorporated into building design can reduce the effects of a vapour phase explosion on occupants.
- **Plant and process design and control** can reduce the risk of fire and explosion by preventing flammable concentrations of vapour developing and coming into contact with sources of ignition. In addition, isolation can stop the explosion from reaching other areas of the plant through pipes or ducts. Pressure-resistant equipment will contain an explosion. Process controls can ensure that concentrations do not exceed the LFL.
- **Control of the amount of material** limits the quantity to be stored in a workroom or working area and requires justification of the need to store any particular quantity of flammable liquid within a workroom/working area. The guiding principle is that only the minimum quantity needed for frequently occurring activities or that required for use during half a day or one shift should be present in the workroom/working area.

  It is recommended that the maximum quantities that may be stored in cabinets and bins are no more than:

  - 50L for extremely/highly flammable liquids and those flammable liquids with a flash point below the maximum ambient temperature of the workroom/working area.
  - 250L for other flammable liquids with a higher flash point of up to 55°C.
- **Prevention of release** by storing flammable liquids in a separate storage area, or in a purpose-made bin or cupboard, keeping containers closed when not in use, using safety containers which have self-closing lids, dispensing liquids over a tray and using non-flammable absorbent material to mop up spills.
- **Segregation and storage of materials** will aim to ensure that flammable substances are stored outside in designated storage areas. If kept inside, they should be segregated from any work that is likely to produce a source of ignition.
- **Hazardous area zoning** classifies areas on the basis of the frequency and duration of the occurrence of an explosive atmosphere. For example, under **DSEAR**, such areas are classified into one of three zones depending on the likelihood that an explosive atmosphere will exist within the area.
- **Inerting** involves the partial or complete substitution of the air in an enclosed space by an inert gas and can be employed very effectively to prevent explosions.
- **Explosion relief** reduces the damage from a vapour phase explosion by either relieving the pressure generated by the explosion by means of vents, panels and bursting discs or by suppressing the explosion through inerting.

- **Control of ignition sources** by removing all the obvious ignition sources from the storage and handling areas for flammable liquids. Typical ignition sources include sparks from electrical equipment or welding and cutting tools, hot surfaces, open flames from heating equipment and smoking materials.
- **Sensing of vapour between LEL and UEL** in order to detect the formation of an explosive atmosphere which can trigger an alarm system or preventive actions such as ventilation or inerting.

Dust Explosions

Industries and Plant with Potential Dust Explosion Hazards

Some dusts are capable of producing explosions which are every bit as destructive as those from gases or vapours. Most organic solids, most metals, and some combustible inorganic salts can form explosive dust clouds. The dust must be **explosible**. (The **ignition energy** can be determined experimentally in test apparatus.) Examples of explosible dusts include materials such as: flour, custard powder, sugar, metal powder, paint powder, coal and grain.

Common processes generating explosible dusts include flour milling, sugar grinding, spray drying of milk and conveyance/storage of whole grains and finely divided materials.

Examples of typical plant where explosive concentrations of dust may exist include:

- Local exhaust ventilation systems.
- Dust-collecting filter units.
- Powder-handling systems.
- Bucket elevators.
- Silos and bins.

The Dust Pentagon

For a fire to start, there must be a fuel, a source of oxygen (usually taken from the surrounding air) and an ignition source. As we saw earlier, this is commonly known as the 'fire triangle'.

In the case of a combustible dust explosion, two more elements are required to be added to the triangle, creating the '**dust explosion pentagon**'.

These two new elements are: confinement of the dust cloud, and dispersion of the dust in air. These elements are created when the fuel (a combustible dust) is dispersed as a cloud within an enclosed area, such as a bin, silo or room.

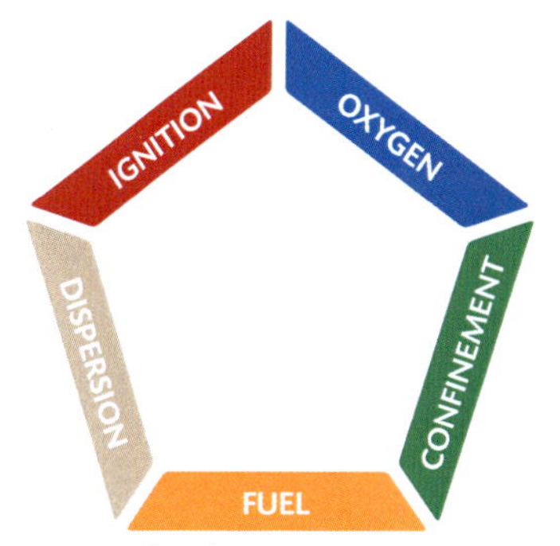

The dust pentagon

In just the same way the fire triangle is useful for understanding fire prevention, so the dust explosion pentagon helps us to understand how to prevent dust explosions (keep the five elements apart).

Mechanisms of Dust Explosions

Other factors that affect the possibility of dust explosions include:

- **Combustible Solid Particulate Size**

  The **dust particles must be of a suitable size**. Solids do not have the divisions of flammability that liquids do. Dusts, however, are finely divided solids and, as with sprays, they have an increased surface area. Generally, finer dusts are more hazardous. The fineness and composition of the dust may also be a factor in determining whether it is explosible, and it is this particle size that we must consider. The smaller the particle, the greater the total exposed surface area of a given mass. The result of putting a source of ignition to a concentration of airborne dust can be the same as discussed with atomised fuel, with very rapid burning.

- **Dispersal**

  The dust particles must be dispersed in air to form a combustible mixture. Dispersal is an important factor leading to secondary explosions, which are caused when dust accumulating on surfaces in the workplace is disturbed by the primary explosion and disperses to form a second dust cloud, which is then ignited by the heat released from the primary explosion.

- **Explosive Concentrations**

  The dust must be mixed with air and have a concentration which lies between the explosive limits:

  - The LEL is generally taken to be about 25g to 50g per cubic metre of air and a cloud will obscure visibility to about one metre - such a concentration would not normally be encountered in the workplace. Note that this dust concentration in air is far greater than the Occupational Exposure Limits (OELs) for substances hazardous to health. A dust cloud will have variable density and tends to become finer over time as the heavier particles settle out.
  - The UEL is ill-defined but generally taken to be about ten times the LEL.

- **Ignition Energy**

  There must be a sufficient source of energy to ignite the dust cloud. It is a fact that explosible dust clouds can be ignited by sparks from electrical equipment such as switches and motors, and in short-circuiting caused by damaged cables. Additionally, electrostatic discharges may initiate dust explosions in industry. For an assessment of the hazard situation in dust-processing installations, knowledge of the **Minimum Ignition Energy (MIE)** is indispensable.

DEFINITION

MINIMUM IGNITION ENERGY - DUST CLOUD

The MIE of a dust cloud is the lowest energy value of a high-voltage capacitor discharge required to ignite the most readily ignitable dust/air mixture at atmospheric pressure and room temperature. The dust concentration and the ignition delay are systematically varied until a minimum value of the ignition energy is found. The tested energy levels are 1,000, 300, 100, 30, 10, 3 and 1mJ.

- **Temperature and Humidity**

  In general, if a process is operating at a higher temperature then the initiation of a dust explosion will require less energy and the combustion reaction will proceed at a greater rate.

  Hot surfaces can initiate ignition of a dust cloud. The minimum ignition temperature is the lowest temperature of a heated vertical surface capable of igniting a mixture of dust and air which comes into contact with it at the optimal concentration.

The humidity of a dust has an effect on its ignition and explosive behaviour. A higher level of humidity will increase the required MIE and may impede the formation of dust clouds. In simple terms, 'damping down the dust' will reduce the likelihood of a dust explosion.

Primary and Secondary Explosions

When dust explosions occur, there is normally an initiating primary explosion inside a piece of equipment such as a silo. This causes a blast wave that ruptures the silo and lofts dust in the general workplace. This dust is then ignited by the escaped burning dust to form a **secondary** explosion.

The concentrations needed for a dust explosion do not usually occur outside of process vessels, so most severe dust explosions start within a piece of equipment (such as mills, mixers, dryers, cyclones, hoppers, silos, etc.). These are known as **primary explosions**. When the equipment is working, dust is generally suspended in air inside the equipment, which can allow dust explosion conditions to occur. This explosion can cause the vessel to rupture.

Secondary explosions are caused when lying dust in the workplace is disturbed by the primary explosion blast wave and forms a second dust cloud, which is then ignited by the heat/fire of the primary explosion. The problem is that small amounts of lying dust occupy very little space, but once disturbed can easily form dangerous clouds. A 1mm layer of dust can give rise to a 5m deep cloud of explosible dust.

There can be a large series of explosions triggered in this manner, with devastating consequences if there is a lot of lying dust that is disturbed. It is important, therefore, to reduce the amount of lying dust to a minimum.

Prevention and Mitigation of Dust Explosions

The prevention and mitigation of dust explosions share many common characteristics of general explosion prevention and protection. This subject will be dealt with in greater detail later in this Learning Outcome.

Here we will simply outline some of the basic principles:

- **Dust explosions are prevented by:**
  - Using appropriately designed and maintained dust-handling equipment so that dust is contained inside safe plant and does not escape out into the wider work environment.
  - Inerting the atmosphere inside handling equipment to reduce oxygen concentrations to a point where the dust will not combust (see earlier section, 'Limits of Flammability').
  - Zone classifying the area where dust is being handled so that appropriate control measures over ignition sources can be exercised.
  - Selecting and maintaining electrical equipment that is suitable for the zone classification of the area.
  - Controlling other ignition sources that might exist within the zoned area, such as hot works and hot surfaces caused by mechanical friction.
  - Preventing the build-up and discharge of static electricity by careful plant and equipment selection and operation (this topic is dealt with in more detail later in the unit).
- **Should dust explosions occur, they are mitigated by:**
  - Ensuring a high level of housekeeping to remove dust deposits outside of handling equipment, so that dust is not available to cause secondary explosions.
  - Making use of isolation mechanisms to isolate explosions inside one item of plant so that combusting dust and gas cannot escape to other parts of the equipment.
  - Using explosion suppression to detect and immediately suppress dust explosions inside handling equipment so the explosion does not have time to fully develop.

- Fitting explosion relief mechanisms, such as explosion relief vents, panels and bursting discs, so that combusting expanding dust and gases are able to escape safely to the atmosphere in a controlled manner.

STUDY QUESTIONS

1. Define what is meant by the terms 'flash point' and 'auto-ignition temperature'.
2. Explain why petrol, with a flash point of approximately −40°C, poses a constant fire risk.
3. Name three types of vapour cloud explosions.
4. Explain the main difference between a Confined Vapour Cloud Explosion (CVCE) and an Unconfined Vapour Cloud Explosion (UVCE).
5. Describe the sequence which triggers a secondary dust explosion.

(Suggested Answers are at the end.)

Behaviour of Structural Materials, Buildings and Building Contents in a Fire

IN THIS SECTION...

- The behaviour of building structures and materials affect the safety of occupants and building contents by influencing the manner and rate of fire spread. The selection of building materials depends on specific use and should consider combustibility, structural strength and products of combustion.
- Common building materials whose fire properties and level of fire resistance we need to understand include steel, concrete and wood.
- The behaviour of and effects of fire on a building depend on the behaviour of common building contents including paper-based materials, fabrics and plastics.

Behaviour of Building Structures and Materials in Fire

Many types of building material are in current use. Different materials respond in different ways to fire and therefore the choice of material will impact upon the manner and rate of fire spread. The selection of appropriate building materials depends on specific use and should consider the following:

- **Combustibility** (how readily it will burn).
- **Structural strength** when subject to heat.
- **Products of combustion** (harmful or otherwise).

Fire Properties of Common Building Materials and Structural Elements

Steel Frames

Steel has a high strength/weight ratio and is commonly used in load-bearing members. Structural members include: columns, beams, portal frames and roofs. Steel can also be used as 'profiled' sheets or lightweight roof members (purlins).

When exposed to fire, unprotected steel will rapidly lose its designed shape and, therefore, its strength. A steel beam reacts to fire by expanding and pushing the columns out, causing the floor slabs to collapse onto the floor below. This floor, not being strong enough to carry the extra load placed upon it, collapses and down everything goes. The steelwork can also spread the heat by conduction, causing the fire to spread.

The choice of material will impact on fire spread

Reinforced Concrete Frames

Almost all concrete used for structural purposes is reinforced. The fire resistance depends on the:

- **Type** of aggregate used. All concrete is likely to 'spall' (break away) when hot, especially when hit with a jet, but the use of lightweight aggregate, or aerated concrete, can minimise this.
- **Thickness** of concrete over the reinforcing rods.

Generally, the fire resistance is good, collapse is not usually sudden and many structures have been reinstated after severe fires.

DEFINITION

SPALLING

A physical process of the breakdown of surface layers of masonry, which crumble into small pebble-like pieces in response to high temperatures or mechanical pressure.

Spalling is caused by heating, mechanical pressure, or both. This heat and/or pressure causes uneven expansion of the materials that make up the masonry.

Pressure created by rapid changes in temperature, such as application of cold water to the heated surface during fire-fighting operations, can also cause spalling. These processes break the bond that holds the solids together and thus cause the masonry to crumble.

Whether reinforced or pre-stressed, the fire resistance of concrete is mainly determined by the protection against an excessive rise in the temperature of the steel (reinforcing rods) offered by the concrete cover. Mild steel loses half its cold strength at about 550°C and high-tensile steel at about 400°C. Structural concrete may then deflect under fire conditions but does not normally collapse suddenly.

Timber

Timber burns, but in a predictable manner. If designed with an adequate factor of safety, there can be a reasonable time lag before failure occurs, particularly if the timber is protected with plasterboard or other coverings.

Applied heat will not cause expansion to stress the structure, nor does wood collapse suddenly.

The fire resistance of timber depends on the following:

- The **thickness** or cross-sectional area of the piece.
- The **tightness** of any joints involved - in general, the fewer joints, the better.
- To some extent, the **type** of wood - generally, denser timber has better resistance. The surface chars but, because conduction is poor, the internal timber still performs structurally.
- **Excess material** (sacrificial timber) is added to exposed beams and columns as necessary. It is consumed by a fire before the structural core is attacked.
- Any **treatment** received - plywood or chipboard sheets may require flame-retardant treatment.

For several years now the use of laminated timber members has been on the increase, simply due to the fact that it can be designed to almost any profile and has excellent fire-resistant properties.

TOPIC FOCUS

The **effects of fire in a workplace on structural materials**:

- **Steel**
  - Heats up in a fire and conducts it throughout its structure.
  - Loses strength at high temperature (>500°C).
  - Expands and buckles losing strength and structure.
  - Regains strength on cooling but properties may have changed.
- **Concrete**
  - Fire-resistant and does not conduct heat.
  - Acts as an insulator at lower temperatures.
  - Will spall and disintegrate as the temperature rises.
  - Reinforcing rods will act as conductors and expand, increasing any spalling.
  - Loses structural strength on cooling.
- **Wood**
  - Will smoulder and char initially.
  - Will burn as the temperature increases.
  - Rate of charring/burning depends on density of the wood and tightness of any joints.
  - Chars and burns from the outside so thicker wood will retain strength longer.
  - Generates smoke and fumes and allows surface propagation.

Brickwork

There are three types of brick in common usage: fired clay, calcium silicate and concrete.

No distinction is made between them in classifying their behaviour in fire when incorporated as a wall. The important features which affect the fire resistances of a wall are:

- Its thickness.
- The applied rendering or plastering, especially if lightweight plaster is used.
- Whether the wall is load-bearing or not.
- The presence of perforations or cavities within the bricks.

Fired clay bricks normally respond better in a fire situation due to the manufacturing process involved, i.e. natural material (clay) and heat. They have already been exposed to greater temperatures, therefore there will be very little movement. However, the above features must still be taken into consideration.

Clay bricks in use on a construction site

Building Blocks

Blocks may be of clay or concrete. Their fire resistance is improved by the application of plaster:

- **Clay blocks** are generally hollow. The greater the thickness and the smaller the voids, the better the fire resistance. Spalling (blistering and exploding) is likely to occur on the face exposed to fire.
- **Concrete blocks** may be made of dense or lightweight aggregates and can be either solid or hollow. All types give high fire resistance with little risk of collapse, so can be used for the walls of a fire compartment.

Building Boards

Boards are combustible if not easily ignitable. Types include:

- Softboard (non-compressed in manufacture), often called 'insulating board'.
- Hardboard of both low and high density - if tempered by impregnation with oils and resin it has high strength and water resistance and is not easily ignitable.
- Plaster boards which retard fire spread until the paper face burns away.
- Asbestos boards (normally in older buildings) which have a high asbestos content and, consequently, have good fire-resistance properties.
- Asbestos cement sheets (again, normally in older buildings) which have a low asbestos content and usually fail by shattering under fire.
- Plywood, block boards; in both types there is variable fire resistance depending on the type of wood, thickness, etc.
- Plastic board of variable resistance depending on surface treatment.

Building Slabs

Slabs are similar to building boards but are much thicker. 'Wood-wool' slabs and compressed straw slabs are combustible and are often treated to give improved resistance. These are usually found as substrates for roofing materials.

Stone

Stone used in building will be of three general types:

- **Granite**: likely to expand rapidly and shatter at 575°C. There is always a risk of spalling, but less so with large blocks.
- **Limestone**: spalling is likely if hit with a jet at high temperature.
- **Sandstone**: generally comes between limestone and granite in behaviour. It is likely to shrink and crack.

Stone has a tendency to crack when subjected to continuous heat or to sudden cooling by a jet.

Glass

Glass is susceptible to breakage so cannot be used as a barrier to fire. Two exceptions are wired glass and copperlight glazing, which give some fire resistance. Other types of glazing include:

- **Armourplate**: this is toughened glass which is incapable of providing fire resistance and will not stand temperatures above 300°C.
- **Double-glazing**: two or even three sheets of glass are mounted within a frame - again they are not fire-resisting and will probably shatter in a fire.

- **Fire-resistant glass blocks** are available which need to meet the following criteria:
  - Mechanical resistance - the glass block wall must stay upright without too much damage following testing.
  - Thermal isolation.
  - Imperviousness against blaze.
  - No flammable emission during testing.
- **Fire-resistant glazing** is a recent development. Clear glasses which incorporate clear intumescent interlayers or laminates will provide fire resistance of up to 90 minutes.

Insulating Material

Most modern materials are non-combustible; unfortunately, in many older buildings, combustible materials (sawdust) have been used. Their location in concealed spaces can aid fire spread considerably.

Lime (Plaster)

Lime is made by heating limestone (calcium carbonate), which is converted to quicklime (calcium oxide). It is then slaked with water to make calcium hydroxide. Lime is a component of plaster and mortar. It is used for plastering internal walls and, if supported by lathing or expanded metal, has good fire resistance.

Paint

Most paints are flammable; a layer of many coats built up over years may be a fire risk. There are also flame-retardant paints and intumescent paints, which bubble up to protect the timber beneath.

Plastics

There are two types of plastic:

- **Thermosetting plastics** are formed by the action of heat and compression; they will not soften and melt when involved in a fire but will decompose.
- **Thermoplastic plastics** are moulded into the required shape by heating. On cooling they remain in that shape. If involved in a fire, they will melt and flow.

Plastics are used primarily in building services and surface fascia. The principal hazards caused are the dripping of molten plastic and from the products of incomplete combustion in the form of smoke.

Level of Fire Resistance

Even when an acceptable standard of fire resistance has been achieved within a building structure, it can be easily compromised through poor selection of decorative materials and furnishings.

Curtains, carpets, furniture and even wallpaper can provide a ready route for fire transfer, as well as smoke generation. Particular care should be taken in their selection to minimise fire spread.

The requirements for fire resistance within buildings are complex and **Approved Document B of the Building Regulations 2010** states the requirements for the 'surface spread of flame' class for the linings of the walls, floors, ceilings, etc. This will depend on the nature of the undertaking and who may be present on the premises (e.g. the public).

Typically, fire doors will have a fire-resistance rating of 30 minutes, while the structural elements of an industrial building that is no more than 5m in height and without a sprinkler system must have 60 minutes' fire resistance.

Fire resistance of structural elements of buildings is often improved by:

- Insulation of steel members, e.g. covering with concrete.
- Separation into compartments (discussed later).
- Over-engineering - thicker concrete over inner steel reinforcement, thicker wood.
- Selection of material type - best mix for concrete, type of wood.
- Treatments - impregnation of timbers with fire retardants and intumescent coatings.
- Surface cladding with non-flammable materials, e.g. plasterboard.

Behaviour of Common Building Contents in Fire

Paper

Paper is used in various forms - wallpaper, books, magazines, etc. Paper will generally scorch, char and then ignite in a fire, producing smoke, water and carbon oxides. Ignition typically takes place around 250°C. It depends on the type of paper and on its form, e.g. shredded paper in a waste bin is easier to ignite than it would be in the form of a book. The two scenarios have very different surface area to volume ratios. Paper is also often coated in waxes or clays which can affect its ignitability.

Paper will ignite at around 250°C

Plastics

Note the previous section on plastics and the two principal types: thermosetting and thermoplastic. Plastics are widely used inside buildings. Expanded cellular plastics (typically urethane foam) are used as wall and ceiling linings as they offer very good insulation properties. If these linings are unprotected, they can present a serious fire hazard. UPVC window frames are also popular in buildings.

Different plastics can behave very differently under fire conditions. Some ignite, some self-extinguish and some smoulder. Unplasticised halogenated plastics (such as UPVC) are naturally flame-retardant. If plasticisers are added, however, this can render them combustible. The products of combustion of plastics are similar to those of paper or wood. In oxygen-deficient atmospheres, large volumes of smoke and carbon dioxide are produced.

Fabrics and Furnishings

Fabrics and furnishings are also commonly found in buildings in the form of seating, curtains, etc. Fabrics can be natural or synthetic (man-made). Natural fabrics tend to smoulder. Synthetic fabrics (which are predominantly thermoplastics) tend to melt and shrink when exposed to a smouldering ignition source, and then ignite and burn rapidly when exposed to open flame.

STUDY QUESTIONS

6. Describe the behaviour of steel construction materials in a fire.
7. Describe how flame-retardant paint protects covered timber.

(Suggested Answers are at the end.)

Fire and Explosion Prevention and Protection

IN THIS SECTION...

- Structural protection can be achieved through the use of compartmentation, openings and voids and fire-stopping.
- Fire/explosion can be a consequence of poor plant and process design and control. Key features of plant design to prevent or minimise these effects include: isolation, pressure-resistant equipment and plant layout.
- Flammable, combustible and incompatible materials should be segregated.
- Hazardous places are classified in terms of zones on the basis of the frequency and duration of the occurrence of an explosive atmosphere.
- In zoned hazardous areas the effects of ignition sources are controlled by exclusion.
- Inerting involves substitution of the air in an enclosed space by an inert gas and can prevent explosions when the explosive or flammable hazard cannot be eliminated by other means.
- Methods of explosion relief include explosion venting, bursting discs, explosion panels and suppression.

Structural Protection

The main features which influence the rate of fire spread are the:

- Distance from the source of fire.
- Materials with which the building is constructed.
- Layout and construction of the building.
- Materials used in decoration and furnishings.

Each factor can be integrated within building design, so providing **passive** protection. We looked at some aspects of this under 'Level of Fire Resistance' earlier.

Fire spread

Openings and Voids

Ceiling and floor voids, as well as openings around pipework and other services, can allow air to feed a fire and also assist in the spread of fire and smoke. Debris should not be allowed to accumulate in voids. Where necessary, such openings should be bonded or fire-stopped with non-combustible material.

Compartmentation

By siting high fire-risk processes or storage of materials at some distance from occupied buildings, the risk of fire spread is greatly reduced. Unfortunately, due to restricted space this is not always achievable so other means must be sought.

The spread of fire and smoke within a building can be prevented by dividing the building into compartments or fire-tight cells using fire-resistant materials.

Fire compartment walls, floors and doors must generally provide a 60-minute resistance to fire, but this may vary depending on the level of risk within the compartment. Walls, floors and doors subdividing such compartments must generally provide a 30-minute resistance. If any services or ducting pass through any compartment wall, floor, ceiling or roof then the joints around the services, etc. must be **fire-stopped** (see later) to prevent the passage of fumes, smoke or gases.

In areas of high risk, the fire compartment would be designed to keep the fire in. High-risk areas are areas where:

- Highly flammable materials are used, transported and stored.
- Toxic fumes are produced.
- Gas cylinders are used or stored.

In areas of high-loss effect, the fire compartment would be designed to keep fire out. High-loss-effect areas are those where:

- Essential records or documents are kept.
- Essential equipment, plant or stock are located.

Fire doors, where fitted, should have effective self-closing devices and be labelled "Fire Door - Keep Shut". Self-closing fire doors may be held open by automatic door-release mechanisms which conform to one of the following:

- Connected into a manually operated electrical fire alarm system incorporating automatic smoke detectors in the vicinity of the door.
- Actuated by independent smoke detectors (not domestic smoke alarms) on each side of the door.

Where such mechanisms are provided, it should be possible to release them manually. The doors should be automatically closed by any of the following:

- The actuation of a smoke-sensitive device on either side of the door.
- A power failure to the door-release mechanism or smoke-sensitive devices.
- The actuation of a fire warning system linked to the door-release mechanisms or a fault in that system. Such fire doors should be labelled with the words "Automatic Fire Door - Keep Clear".

Fire-Stopping

Very often, mechanical and electrical services breach compartment walls and floors and, where gaps around services have not been adequately fire-stopped, the insulation may fail, leading to a rise in temperature along the conductive materials and the danger that conducted heat may propagate fires elsewhere in the building. Smoke and heat may also pass through these inadequately insulated openings, endangering occupants and fire-fighters attempting to extinguish the fire.

Fire-stopping in fire-resisting separating compartments plays a critical role in containing a fire at its source and can be described as preventing the spread of smoke and flame by placing obstructions across air passageways. Ventilation ducts and gaps around doors must have the facility to be stopped in the event of a fire. This can be achieved by the use of baffles, self-closing doors and intumescent material which expands when heated, sealing the opening.

Approved Document B of the Building Regulations and BS EN 1366-4 *Fire resistance tests for service installations - Linear joint seals* give advice on the types of materials to be used and testing arrangements for CE-marked products.

Key Features of Plant Design and Process Control

Fire or explosion can be a consequence of poor plant and process design and control, where flammable concentrations of vapour or dust are allowed to develop and come into contact with sources of ignition. So we need to be aware of the key features of plant design and process control that should be taken into account in order to prevent or minimise the effects of such occurrences.

Plant Design

Isolation

By isolating an explosion, the amount of damage caused can be reduced in a number of ways. Isolation stops the explosion from reaching other areas of the plant through pipes or ducts. It also stops the possibility of jets of flame that can occur at the end of long pipes and can prevent the pressure from the primary explosion causing secondary explosions.

There are two different ways in which isolation can be achieved:

- Passive (activated by the explosion itself).
- Active (which requires tripping by a sensor for activation).

Passive systems are simpler and more reliable.

Passive isolation involves designing and building the dust-handling equipment so that it is not possible for combusting dust and gas to pass from one vessel into other parts of the equipment. This is often done by designing a choke into the handling system. For example, a screw-feed conveyor with a baffle plate part way down its length, or a rotary valve at the bottom of a cyclone or dust collector.

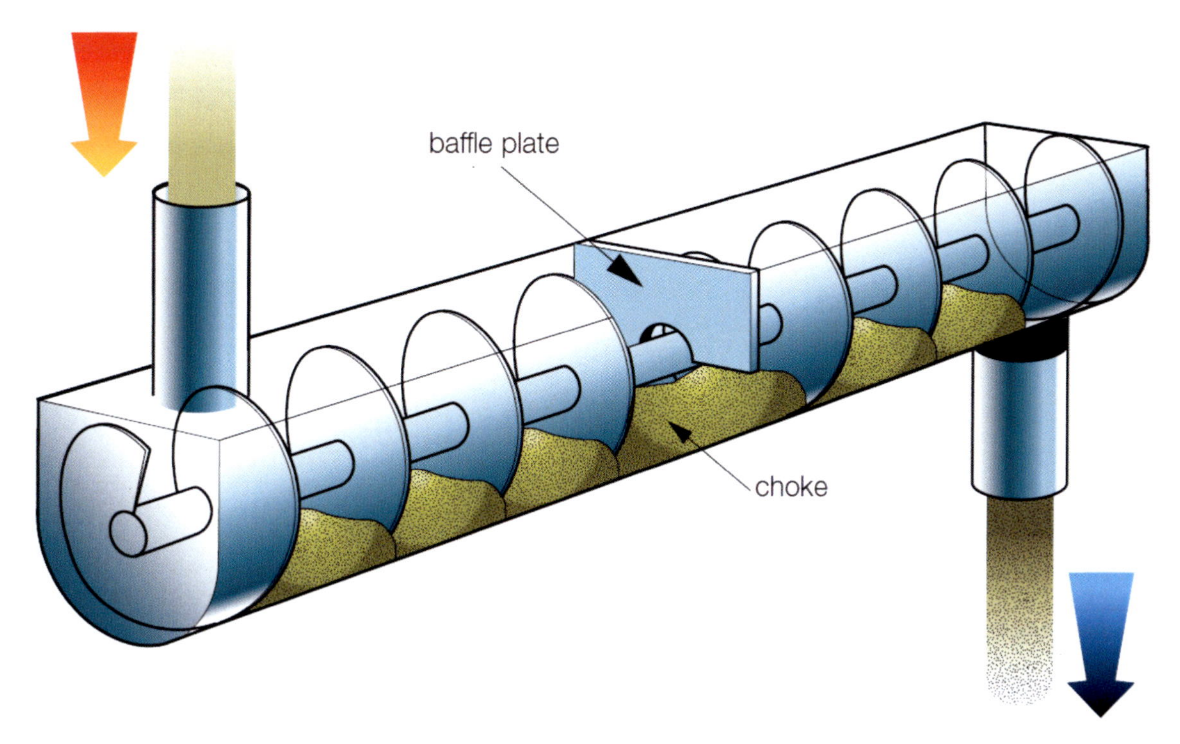

Use of a screw conveyor as an explosion choke

Source: HSG103 *Safe handling of combustible dusts: Precautions against explosions*, 2nd ed., HSE, 2003 (www.hse.gov.uk/pubns/priced/hsg103.pdf)

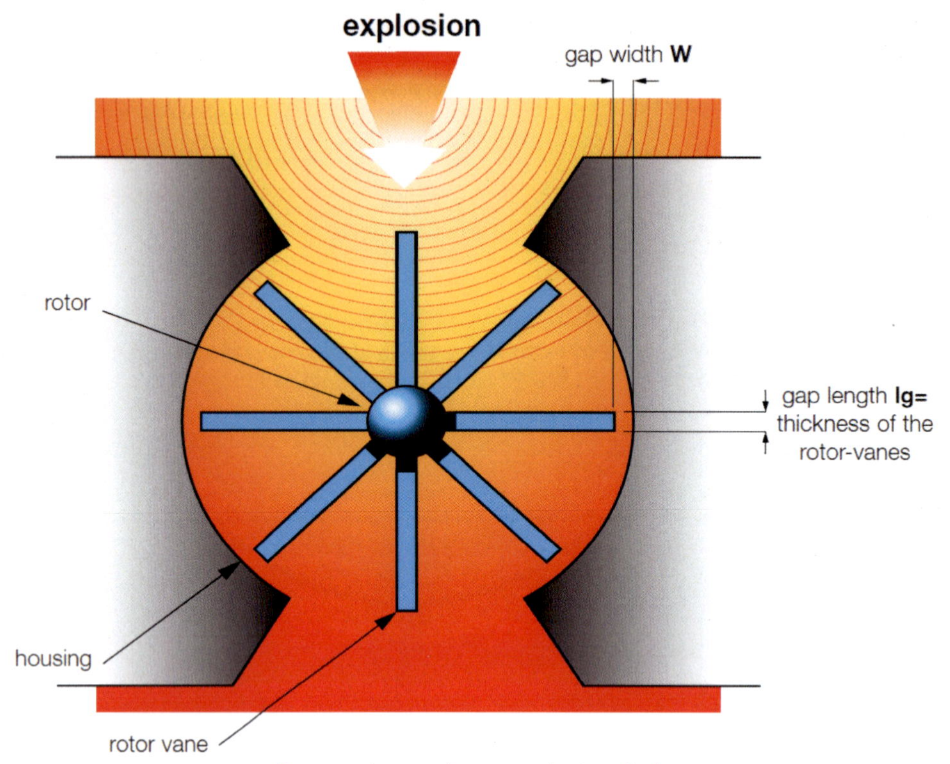

Rotary valve used as an explosion choke

Source: HSG103 *Safe handling of combustible dusts: Precautions against explosions*, 2nd ed., HSE, 2003 (www.hse.gov.uk/pubns/priced/hsg103.pdf)

Active-isolation systems rely on very fast early detection of the pressure rise within a vessel as a result of the explosion initiation. This then triggers the isolation mechanism to activate. The mechanism often involves the closing of fast-acting valves or may involve the release of a suppressant such as dry powder in inert gas into adjacent parts of the dust-handling equipment.

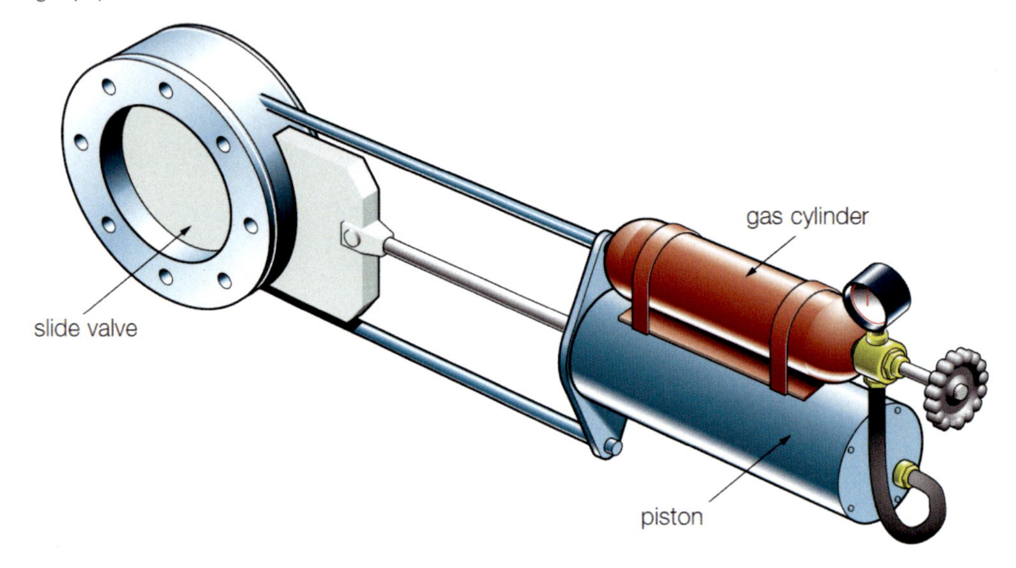

High-speed isolation valve

Source: HSG103 *Safe handling of combustible dusts: Precautions against explosions*, 2nd ed., HSE, 2003 (www.hse.gov.uk/pubns/priced/hsg103.pdf)

Pressure-Resistant Equipment

Perhaps the best way to contain the primary dust explosion is to have process equipment that is strong enough to withstand the explosion. Most vessels that may be at risk of dust explosions are designed so that, in an explosion, they will distort but not rupture.

Plant Layout

Buildings that could be in danger of dust explosions must be designed along the same lines as any plant which has the danger of explosion. The following features should be considered:

- Wherever possible, they should be isolated from other buildings.
- Buildings should preferably be one storey high.
- If inside, the vulnerable area of the building must be reinforced.
- The rest of the areas of the plant must be protected (e.g. a blast wall may be needed).
- Sufficient venting to avoid structural damage from overpressure must be provided.
- Hot gases from the explosion must be vented to the outside atmosphere to prevent secondary fires.

Other features are safe escape routes in case of an explosion and fire, fire-resistant construction materials, fire-resistant doors and good electrical insulation.

Process Controls

One of the basic control methods used to prevent fire or explosion during the processing and handling of various vapour and gases is to keep the mixture of fuel and air outside of the relevant flammability limits. If gas or vapour concentrations can be kept below the LFL (LEL) then there is no fire or explosion risk. Natural or forced ventilation can therefore be used as an effective fire prevention method. Similarly, in areas where high levels of gas or vapour are likely to occur (such as inside process vessels), if the gas or vapour concentration is kept above the UFL (UEL) then ignition, combustion and explosion cannot occur.

The use of gas/vapour detectors and alarm systems can be used to measure the vapour/gas in air concentrations to give warning of mixtures that are within the flammable range. These systems can also be connected to other control mechanisms to trigger appropriate action.

Process controls can be used to avoid creating an explosive dust cloud by:

- Adding inert gas to the atmosphere.
- Ensuring that the dust is outside of the combustible concentration limits.
- Adding inert dust.

Segregation and Storage of Flammable, Combustible and Incompatible Materials

Flammable, combustible and incompatible substances should be stored outside in designated storage areas with an appropriate level of segregation and separation. Where these materials have to be stored inside, they should be stored in rooms or cabinets with an appropriate level of fire resistance. Separation and segregation of sensitive materials will be necessary. **Full details on the safe storage of dangerous substances can be found later in the unit**.

Hazardous Area Zoning

TOPIC FOCUS

DSEAR set minimum requirements for the protection of workers from fire and explosion risks related to dangerous substances and potentially explosive atmospheres, from gases under pressure,and from substances corrosive to metals. **DSEAR** requires employers to control the risks to the safety of employees and others from these hazards.

Hazardous places are classified in terms of zones on the basis of the frequency and duration of the occurrence of an explosive atmosphere. Schedule 2 of the **DSEAR** describes the different zones:

Flammable vapours, gases and mists:

- **Zone 0**

  A place in which an explosive atmosphere consisting of a mixture with air of dangerous substances in the form of gas, vapour or mist is present **continuously or for long periods of time or frequently**.

- **Zone 1**

  A place in which an explosive atmosphere consisting of a mixture with air of dangerous substances in the form of gas, vapour or mist is **likely to occur in normal operation occasionally**.

- **Zone 2**

  A place in which an explosive atmosphere consisting of a mixture with air of dangerous substances in the form of gas, vapour or mist is **not likely to occur in normal operation but, if it does occur, will persist for a short period only**.

Combustible dusts:

- **Zone 20**

  A place in which an explosive atmosphere in the form of a cloud of combustible dust in air is present **continuously, or for long periods or frequently**.

- **Zone 21**

  A place in which an explosive atmosphere in the form of a cloud of combustible dust in air is **likely to occur in normal operation occasionally**.

- **Zone 22**

  A place in which an explosive atmosphere in the form of a cloud of combustible dust in air is **not likely to occur in normal operation but, if it does occur, will persist for a short period only**.

Determining which zone is suitable for different circumstances and where the boundaries between zones should be drawn is a complex process which should involve expertise from a number of fields such as safety, production and chemical engineering.

Once the zone boundaries are established, equipment can be designed so that it is suitable for the intended area. This is an important stage in avoiding an explosion in the first place. The **Electricity at Work Regulations 1989** state that any electrical equipment which may be exposed to any flammable substance, including dusts, vapours or gases, should be of such construction or as necessary protected so as to prevent danger arising from that exposure. (This topic is dealt with in more depth later in this unit.)

Exclusion of Ignition Sources

MORE...

Additional information on hazardous area classification and control of ignition sources is available from the HSE at:

www.hse.gov.uk/comah/sragtech/techmeasareaclas.htm

Controls for the exclusion of ignition sources in zoned hazardous areas will include:

- Selection of equipment that is designed to be safe for use in the area.
- Inspection and maintenance regimes to minimise the likelihood of faulty equipment giving rise to an ignition source.
- Ensuring that any portable/mobile equipment brought into the area is either suitable, or that the area is made safe before it is brought in. (Entry is controlled by permit-to-work systems.)
- Prohibition of open flames, e.g. smoking.
- Control of static discharge, e.g. by earthing.
- Ensuring heating equipment in such areas cannot act as an ignition source.
- Good housekeeping - not allowing combustibles to build up.
- Segregation of incompatible chemicals that could react to produce heat/flame/sparks.

Inerting

Inerting involves the partial or complete substitution of the air in an enclosed space by an inert gas such as nitrogen or carbon dioxide. It can be employed very effectively to prevent explosions, but is usually only used when the explosive or flammable hazard cannot be eliminated by other means, such as the use of non-flammable materials or the methods considered previously.

Inerting as a method of explosion control and prevention of fire spread is used in enclosed spaces and plant. Plant that is substantially open will have varying oxygen concentrations and so cannot be successfully inerted. Inerting is used within reactor systems where the possibility of a flammable atmosphere exists, and within tanks where a material may be stored at temperatures above its flash point.

Methods of Explosion Relief

Explosion relief is a form of explosion **mitigation**. In other words, it is a method for dealing with the effects of an explosion that has not been successfully prevented.

Explosion Venting

Explosion venting works by directing the expanding/combusting gas/dust cloud in a safe direction (away from the operator) to prevent excess pressure build inside vessels and handling equipment, that would otherwise cause catastrophic failure. Vents must be designed to allow sufficient outflow of burnt gas/dust and air to relieve the pressure generated by the explosion. Care must be taken not to vent toxic dusts into the atmosphere, but venting to a sealed area may be possible. Vents are sealed with materials that are designed to fail when internal pressure builds rapidly - in effect, vent seals are a weak link in the containment.

There are a number of designs for venting, including bursting discs, explosion panels and quench tubes:

- **Bursting Discs**

  Bursting discs are normally circular in shape and are fitted within or at the end of vent pipework. Bursting discs are commonly used for overpressure protection, or sometimes to protect against excessive vacuum. They are the weakest point in the system - designed to fail and so avoid mechanical damage to the rest of the system. They usually consist of a thin metal disc, chosen so that it will fail at a set pressure. They are quite cheap but, of course, are a single use item; once they fail, they must be replaced.

- **Explosion Panels**

  These are larger, square or rectangular panels, designed for higher rates of pressure rise (e.g. during an explosion) and for larger capacities of gas to vent. Panels are normally fitted into the sides or top of process vessels, such as dust collectors. They may consist of a metal hinged panel with a plastic membrane underneath. In the event of an explosion, the plastic membrane will fail and the hinged panel will be thrown open by the escaping cloud of burning, expanding gas/dust.

- **Quench Tubes**

  These are mechanisms that are placed over the explosion vent so that the escaping cloud of fast moving burning gas/dust can be slowed and extinguished as it escapes the process vessel. In this way, the significant hazards associated with having a jet flame escaping from a process vessel are diminished.

In all cases of explosion relief venting, care must be taken to ensure that:

- the areas that the burning/expanding gas/dust cloud will vent into are safe to receive such an event and, in particular, are free of people and equipment that might otherwise be injured/damaged or set alight; and
- the vents themselves are not obstructed which would then prevent their proper function and might result in an overpressure event in the vessel which might then rupture.

Suppression

Explosion suppression makes use of an extinguishing suppressant, such as dry powder, suspended in an inert gas, such as CO_2 or argon, to extinguish the combustion reaction inside process vessels or handling equipment before the explosion has time to fully develop and produce a significant blast wave. This requires very rapid detection of the pressure rise inside the vessel that indicates that an explosion is taking place; followed by very rapid activation of the suppressant into the vessel. In effect, the explosion is stopped before it has time to properly get going. Explosion-suppression systems usually make use of state-of-the-art electronic technology to give detection and activation times measured in a few tens of milliseconds.

STUDY QUESTIONS

8. How can the spread of fire within a building be minimised?
9. Describe what is meant by 'fire-stopping' and how this may be achieved.
10. Identify four features that should be considered for buildings that could be at risk from dust explosions.
11. Describe what is meant by Zone 0 and Zone 20 in regard to hazardous area zoning.
12. When considering explosion venting, what factor, other than the explosion itself, must be taken into account?
13. What is the basic principle of operation of a bursting disc?

(Suggested Answers are at the end.)

Summary

Flammable and Explosive Materials and the Mechanisms by which they Ignite

We have:

- Established that:
  - To start a fire, the following three elements of the fire triangle are needed: fuel, oxygen and an ignition source (heat).
  - Suitable ignition sources must be present such as naked flames, hot surfaces or sparks.
  - Combustion can be divided into the following five stages: induction, ignition, growth, steady state and decay.
- Examined the relevant properties of solids, liquids and gases with respect to their influence on combustion.
- Defined flash point, fire point, auto-ignition temperature, vapour density and limits of flammability with examples of the importance of those properties in relation to the initiation and propagation of fire and explosion.
- Discussed the causes and effects of:
  - **Unconfined Vapour Cloud Explosions (UVCEs)** resulting from the release of a considerable quantity of flammable gas or vapour into the atmosphere, and its subsequent ignition.
  - **Boiling Liquid Expanding Vapour Explosions (BLEVEs)** that involves a sudden release of vapour, containing liquid droplets, owing to the failure of a storage vessel.
  - **Confined Vapour Cloud Explosions (CVCEs)** which occur when a flammable vapour cloud is ignited in a container or confined space.
- Examined the mechanisms of explosions and fire spread including:
  - How a fire or explosion occurs.
  - The stages of combustion, including: ignition, growth, steady state and decay.
  - UVCEs, CVCEs and BLEVEs.
- Explained the effects of atomisation/particle size and oxygen content on the likelihood and severity of a fire or explosion.
- Explained how failure of control measures coupled with the physico-chemical properties of flammable materials can bring about an explosion.
- Examined the process of oxidisation and the effects of oxidising substances on fire and explosion mechanisms.
- Examined how flammable atmospheres arise and where they are found.
- Discussed control measures for entering flammable atmospheres, including purging, to keep flammable atmospheres below Lower Explosion Limits (LEL).
- Covered the principles of selection of electrical equipment for use in flammable/explosive atmospheres.
- Examined the prevention and mitigation of vapour phase explosions through structural protection, plant design and process control, segregation and storage of materials, hazardous area zoning, inerting, explosion relief, control of amount of material, prevention of release, control of ignition sources and sensing of vapour between LEL and UEL.

- Considered:
  - Examples of industries/plant with potential dust explosion hazards.
  - The dust pentagon.
  - Mechanisms of dust explosions including the importance of:
    - Combustible solid particle size.
    - Dispersal.
    - Explosive concentrations.
    - Ignition, energy, temperature and humidity.
  - Primary and secondary explosions.
  - Prevention and mitigation of dust explosions including relevant hazardous places zoning (**DSEAR** Schedule 2).

Behaviour of Structural Materials, Buildings and Building Contents in a Fire

We have:

- Studied the behaviour of:
  - Building structures and materials in fire:
    - Fire properties of steel, concrete, wood, other materials, and level of fire resistance.
  - Common building contents in fire:
    - Paper-based, fabrics, plastics.

Fire and Explosion Prevention and Protection

We have:

- Examined structural protection issues such as openings, voids, compartmentation and fire-stopping.
- Studied key features of plant design and process control:
  - Isolating an explosion (passive or active), to reduce the amount of damage.
  - Pressure-resistant equipment designed to be strong enough to withstand the explosion.
  - Plant layout - buildings that could be in danger of dust explosions must be designed along the same lines as any plant which has the danger of explosion.
  - Process controls can be used to avoid creating an explosive dust cloud by:
    - Adding inert gas to the atmosphere.
    - Ensuring that the dust is outside of the combustible concentration limits.
    - Adding inert dust.
- Noted the requirement to segregate flammable, combustible and incompatible materials.
- Studied hazardous area zoning and the requirement for hazardous places to be classified in terms of zones on the basis of the frequency and duration of the occurrence of an explosive atmosphere. In zoned hazardous areas, the effects of ignition sources should be controlled by various means of exclusion.
- Noted that inerting can be used to prevent explosions and fire spread.

- Studied methods of explosion relief and in particular:
  - Vents must be designed to allow sufficient outflow of burnt dust and air to relieve pressure generated by the explosion.
  - Bursting discs are commonly used for overpressure protection, or sometimes to protect against excessive vacuum.
  - Explosion panels are designed for higher rates of pressure rise.
  - Ensuring the pressure rise from the developing explosion is detected by an appropriate sensor and triggers the discharge of an extinguishing agent.
  - Suppression, which makes use of an extinguishing suppressant, such as dry powder, suspended in an inert gas, such as CO_2 or argon, to extinguish the combustion reaction inside process vessels or handling equipment before the explosion has time to fully develop and produce a significant blast wave.

Learning Outcome 11.4

NEBOSH National Diploma for Occupational Health and Safety Management Professionals

ASSESSMENT CRITERIA

- Summarise the main legal requirements relating to fire, the considerations for fire risk assessment, methods for fire prevention and detection, types of fire-fighting equipment, means of escape and emergency evacuation procedures.

LEARNING OBJECTIVES

Once you've studied this Learning Outcome, you should be able to:

- Outline the main legal requirements for fire safety in the workplace.
- Explain the processes involved in the identification of hazards and the assessment of risk from fire.
- Describe common fire detection and alarm systems and procedures.
- Outline the factors to be considered when selecting fixed and portable fire-fighting equipment for the various types of fire.
- Outline the factors to be considered in providing and maintaining the means of escape.
- Explain the purpose of, and essential requirements for, emergency evacuation procedures.

Contents

Legal Requirements

IN THIS SECTION...

- The **Regulatory Reform (Fire Safety) Order 2005 (RRFSO)** (or alternative related local Statutory Instruments such as the **Fire Safety (Scotland) Regulations 2006** and the **Fire Safety Regulations (Northern Ireland) 2010)** is enforced by serving **Alterations Notices**, **Enforcement Notices** and **Prohibition Notices** or through **prosecution** in a criminal court.
- The regulation of fire safety is predominantly the responsibility of the fire and rescue authority but is also shared with the enforcing authorities for health and safety at work such as the HSE and local authority.
- The **RRFSO** requires the employer to:
  - Carry out a fire safety risk assessment.
  - Take steps to remove or reduce identified risks.
  - Meet requirements on the means of escape and fire-fighting equipment.
  - Take measures to mitigate the spread of fire.
- The purpose of the **Building Regulations 2010 Approved Document B** and other national legislation such as the **Building Regulations (Northern Ireland) 2012 Technical Booklet E** and the **Building Regulations (Scotland) 2004 (Standard Technical Handbook 2019: non-domestic, Part 2)** is to secure reasonable standards of health and safety for people in and around buildings. Supporting guidance on how to implement the requirements in practice is set out in Approved Document B.

Regulatory Powers of a Fire Authority with Respect to Fire Safety

The main means of enforcement of the **Regulatory Reform (Fire Safety) Order 2005 (RRFSO)** are:

- The serving of **Alterations Notices**, **Enforcement Notices** and **Prohibition Notices** upon responsible persons and others who may have control of premises.
- **Prosecution** in a criminal court, where appropriate, due to the nature and severity of the offence.

Alterations Notices

Alterations Notices are issued on the premises at the discretion of the enforcing authority.

They are issued if the premises constitute a serious, or may constitute a serious, risk if changes were to be made to them or their use.

If a Notice is issued, the person receiving it must inform the enforcing authority of any proposed changes.

Enforcement Notices

Where a fire authority, or other enforcing authority, is of the opinion that there has been a failure to comply with any provision of the **RRFSO**, an Enforcement Notice may be served on the 'responsible person' (or other person with control over the premises - we'll consider this later).

The Notice will specify the provisions which have not been complied with and require the responsible person to take appropriate remedial actions within a specified period of time.

Prohibition Notices

Where a fire authority, or other enforcing authority, is of the opinion that there are serious risks to the ability of people to escape from premises in the event of a fire, a Prohibition Notice may be served on the 'responsible person' (or other person with control over the premises).

The Notice will specify the matters which give rise to the risk and either prohibit, or place restrictions on, the use of the premises until appropriate actions have been taken to adequately control the risk.

A Prohibition Notice may include steps to be taken to remove the risk from fire and will come into effect immediately. The person on whom the Notice is served may appeal against the notice to a Magistrates' Court within 21 days of it being served although the notice is **not** suspended pending the hearing of such an appeal.

Powers of Inspectors

Under the **RRFSO**, an inspector has the power, at any reasonable time, to:

- Enter and inspect any premises he believes necessary (without the use of force).
- Make all necessary enquiries in order to identify the responsible person and/or to ascertain compliance with the provisions of the **RRFSO**.
- Request any records or plans required for examination or inspection and inspect and take copies of the records.
- Require any person having responsibilities for the premises to give him such facilities and assistance as are necessary to enable him to conduct his duties.
- Take samples of any articles or substances found in any premises for the purpose of ascertaining their fire resistance or flammability.
- Where it is believed that an article or substance has caused, or is likely to cause, danger to the safety of relevant persons, dismantle it or subject it to any process or test.

It is an offence to obstruct inspectors in carrying out their duties or to fail to comply with any requirements that inspectors may impose.

Dual Enforcement by the HSE and Fire Authority

The **RRFSO** serves to link fire safety with broader workplace safety by employing a similar risk assessment approach. Consequently, the regulation of fire safety is predominantly the responsibility of the fire and rescue authority but is also shared with the enforcing authorities for health and safety at work such as the HSE and local authority.

TOPIC FOCUS

Responsibilities for enforcement:

- **The fire authority:**
  - For the majority of workplaces and premises.
- **The Health and Safety Executive (HSE):**
  - For nuclear installations.
  - For ships, while under construction or repair.
  - For construction sites.
- **The local authority:**
  - For sports grounds.

Each of these enforcement agencies has the power to serve Alterations Notices, Enforcement Notices and Prohibition Notices and to bring prosecutions.

In addition to the more obvious emergency role, the fire authority is responsible for the investigation of fires and, on the basis of information gained, the development and implementation of fire prevention and protection policies.

An "**Authorised Officer**" is an employee of a fire and rescue authority who is authorised in writing by the authority.

The powers of an Authorised Officer under the **Fire and Rescue Services Act 2004** are different from those of an inspector under the **RRFSO** and vary depending on whether the officer is acting:

- in an emergency situation, e.g. fire-fighting; or
- in an investigatory capacity, e.g. trying to establish the cause and progression of a fire.

Requirements of the Regulatory Reform (Fire Safety) Order 2005

HINTS AND TIPS

It is permissible to refer to equivalent legislation for Scotland or Northern Ireland in answering NEBOSH questions.

However, you must ensure that such references are clearly stated, e.g. "In Scotland..."

The **RRFSO** requires the employer to:

- Carry out a fire safety risk assessment.
- Take steps to remove or reduce identified risks.
- Meet requirements on the means of escape and fire-fighting equipment, including the maintenance of common fire protection systems.
- Take measures to mitigate the spread of fire.

The **RRFSO** simplified and reformed much of the previous legislation relating to fire safety in non-domestic premises and places a greater emphasis on fire prevention.

The **RRFSO** extends only to England and Wales, with Scotland covered by the **Fire Safety (Scotland) Regulations 2006** and Northern Ireland by the **Fire and Rescue Services (Northern Ireland) Order 2006**.

The principal duties are:

- To ensure, so far as is reasonably practicable, the safety of **employees**.
- In relation to **non-employees**, to take such fire precautions as may reasonably be required in the circumstances to ensure that premises are safe.
- To carry out a **risk assessment**.

Duties of the Responsible Person under the RRFSO

DEFINITIONS

The **RRFSO** introduces two important terms:

RESPONSIBLE PERSON

In a workplace this is the employer, if the workplace is to any extent under their control, and any other person who may have control of any part of the premises, e.g. the occupier or owner. In all other premises, the person or people in control of the premises will be responsible. If there is more than one responsible person in any type of premises, all must take all reasonable steps to work with each other.

RELEVANT PERSONS

Any person who is, or may be, lawfully on the premises, and anyone in the immediate vicinity who is at risk from a fire on the premises. (Note: this does not include fire-fighters when carrying out fire-fighting or other emergency duties.)

These duties can be summarised as follows:

- Carry out a **risk assessment** in order to identify the general fire precautions that are required.
- Take **general fire precautions** to ensure that the premises are safe including:
  - Reducing the risk of fire and the risk of spread of fire.
  - Providing adequate means of escape.
  - Ensuring that the means of escape can be safely and effectively used at all material times.
  - Providing means of fighting fire.
  - Providing means for detecting fire and providing warning in case of fire.
  - Action to be taken in the event of fire, including:
    - The instruction and training of employees.
    - Measures to mitigate the effects of the fire.
- Appoint one or more **competent persons** to assist in the implementation of appropriate preventive and protective measures.

Persons responsible under RRFSO must ensure effective means of escape

- Implement **appropriate arrangements** for the effective:
  - planning,
  - organisation,
  - control,
  - monitoring, and
  - review,

  of the preventive and protective measures.

  These arrangements must be recorded if five or more people are employed or if a Licence (for licensed premises such as a pub) or Alterations Notice is in force.

- Apply the following **principles of prevention** when implementing any preventive and protective measures:
  - Avoid risks.
  - Evaluate risks which cannot be avoided.
  - Combat risks at source.
  - Adapt to technical progress.
  - Replace the dangerous with the non-dangerous or less dangerous.
  - Develop a coherent overall prevention policy.
  - Give priority to collective protective measures over individual protective measures.
  - Give appropriate instructions to employees.
- Provide employees with **information** on the fire risks, the preventive and protective measures, and emergency procedures.
- Provide the **employer** of any **other persons working on the premises** with **information** on the fire risks, the preventive and protective measures, and emergency procedures.
- Ensure that the premises, and any facilities or equipment provided in relation to fire safety, are **maintained**.
- Eliminate or reduce risks from dangerous substances.
- Ensure that appropriate equipment for **detecting fire**, **raising the alarm** and **fighting fire** is provided.
- Ensure that **emergency routes and exits**:
  - Are kept clear at all times.
  - Lead as directly as possible to a place of safety.
  - Are adequate for the use and size of the premises and also the maximum number of persons who may be present at any one time.
  - Have doors that open in the direction of escape.
  - Do not involve sliding or revolving doors.
  - Have doors that are capable of being easily and immediately opened by any person in an emergency.

Responsible person making arrangements with management

  - Are indicated by appropriate signs.
  - Are provided with adequate emergency lighting.
- Ensure that employees are provided with adequate safety **training**.
- Establish appropriate procedures to be followed in the event of **serious and imminent danger** including competent people to implement evacuation procedures.
- Ensure that additional emergency measures in respect of **dangerous substances** are in place.

Fire Safety Act 2021

This Act was brought into force in England on 16 May 2022 by the **Fire Safety Act 2021 (Commencement) (England) Regulations 2022**.

The Act requires that where a building contains two or more sets of domestic premises, the **Regulatory Reform (Fire Safety) Order 2005 (RRFSO)** applies to the building's structure and external walls and any common parts, and all doors between domestic premises and common parts such as flat entrance doors. The Act clarifies the **RRFSO** to make it clear that Responsible Persons must consider these parts when conducting fire risk assessments. Fire and rescue authorities can issue enforcement notices if they decide that Responsible Persons or Dutyholders have failed to comply with any provisions of the **RRFSO**. They can prosecute or serve alterations or prohibition notices if they identify that failing to comply with those provisions puts people at risk of death or injury from fire.

The Government has produced a Fire Risk Assessment Prioritisation Tool that takes Responsible Persons through a series of specific questions, which are each carefully scored to assist them to determine the priority of their buildings for the purpose of reviewing their fire risk assessments. The tool does this by allocating each building to one of five priority tiers.

The tool is part of the package of risk-based guidance that when complied with, will help a Responsible Person show that they have undertaken appropriate steps towards establishing compliance with the **RRFSO.**

The Fire Risk Assessment Prioritisation Tool allocates buildings into five priority categories, based on their weighted numerical scores, to help with the prioritisation of fire risk assessments:

- Tier 1 – very high priority.
- Tier 2 – high priority.
- Tier 3 – medium priority.
- Tier 4 – low priority.
- Tier 5 – very low priority.

Fire Safety (England) Regulations 2022

The Government has released a series of guidance notes for the **Fire Safety (England) Regulations 2022**, which will come into force on 23 January 2023. The Regulations are laid under article 24 of the **Regulatory Reform (Fire Safety) Order 2005**. Regulations impose requirements on Responsible Persons or others, including building owners and building managers, in relation to mitigating the risk to residents for specific premises.

These Regulations will make it a requirement in law for Responsible Persons of high-rise blocks of flats to provide information to Fire and Rescue Services to assist them to plan and, if needed, provide an effective operational response.

In high-rise residential buildings, Responsible Persons will be required to:

- **Building Plans:** provide their local Fire and Rescue Service with up-to-date electronic building floor plans and to place a hard copy of these plans in a secure information box on site.
- **External Wall Systems:** provide to their local Fire and Rescue Service information about the design and materials of a high-rise building's external wall system and to inform the Fire and Rescue Service of any material changes to these walls.
- **Lifts and Other Key Fire-Fighting Equipment:** undertake monthly checks on the operation of lifts intended for use by fire-fighters, and evacuation lifts in their building and check the functionality of other key pieces of fire-fighting equipment.
- **Information Boxes:** install and maintain a secure information box in their building.
- **Wayfinding Signage:** install signage visible in low light or smoky conditions that identifies flat and floor numbers in the stairwells of relevant buildings.

In residential buildings with storeys over 11 metres in height, Responsible Persons will be required to:

- **Fire Doors:** undertake annual checks of flat entrance doors and quarterly checks of all fire doors in the common parts.

In all multi-occupied residential buildings with two or more sets of domestic premises, Responsible Persons will be required to:

- **Fire Safety Instructions:** provide relevant fire safety instructions to their residents.
- **Fire Door Information:** provide residents with information relating to the importance of fire doors in fire safety.

Guidance notes have been issued for each requirement.

Purpose of the Building Regulations 2010: Approved Document B - Fire Safety

MORE...

Approved Document B (Fire Safety) relates to the **Building Regulations 2010** - the latest online version is available at:

www.gov.uk/government/publications/fire-safety-approved-document-b

Building Regulations are made under the **Building Act 1984 (as amended)** and apply to relevant building work in England and Wales - separate legislation applies in Scotland and Northern Ireland.

The purpose of the **Building Regulations 2010** is to secure reasonable standards of health and safety for people in and around buildings. Supporting guidance on how to implement the requirements of Part B in practice is set out in Approved Document B.

The fire safety aspects of the **Building Regulations** are set out in Schedule 1 of Part B and take the form of functional (i.e. performance-based) requirements set out in terms of what is reasonable, adequate or appropriate.

Recently, the Government made a number of changes which are specifically designed to make the guidance more accessible and easier to use, such as splitting the guidance into *Approved Document B - Volume 1: Dwellings* and *Approved Document B - Volume 2: Buildings other than Dwellings*. These new documents provide guidance on the Regulations and introduce further design freedoms and flexibilities to those managing buildings.

Generally speaking, both volumes contain the following information:

- Means of warning and escape.
- Internal fire spread (linings).
- Internal fire spread (structure).
- External fire spread.
- Access and facilities for fire and rescue services.

Volume 2 of *Approved Document B* deals with buildings including offices, public buildings, multi-storey and industrial buildings as well as flats, and because of the nature of these types of buildings, having multi-occupancy and a greater risk of fire, the regulatory requirements are much more thorough than for domestic premises.

Building Safety Act 2022

The **Building Safety Act 2022** came into force in April 2022 and created the Building Safety Regulator (BSR). This role, fulfilled by the HSE, will oversee safety standards in all buildings.

There are a great many provisions within the Act that will not come into force until secondary legislation is introduced.

The BSR (sections 2-30) will have three main functions:

- Overseeing the safety and standards of all buildings.
- Helping and encouraging the built environment industry and building control professionals to improve their competence.
- Leading implementation of the new regulatory framework for high-rise buildings.

Sections 22-24 give the BSR the power to enforce compliance with building regulations. Prosecutions can be of individuals, of corporate bodies (who have contravened building regulations) and other individuals/bodies who have defined roles and duties under the Act.

A Construction Products Regulator will also be established with powers to remove dangerous products from the market and act against those who do not comply with the regulations.

Dutyholders must be competent. This means that the individuals appointed to hold these regulated positions must possess the necessary knowledge, skills and experience in order to carry out their roles.

There is a requirement under the Act for a 'golden thread' of information to demonstrate that a building is compliant with the appropriate building regulations, during the construction phase. It will be the duty of the people responsible for a building to put in place and maintain a golden thread of information. The golden thread needs to be created before building work starts and the golden thread must be kept updated throughout the design and construction process.

All occupied high-rise residential buildings will need to be registered with the regulator. Those responsible for high-rise residential buildings (a new role known as the 'Accountable Person') will need to apply to the regulator for a Building Assessment Certificate (sections 76 to 82). This requirement is being phased in as part of transitional arrangements over five years.

Those managing or owning higher-risk buildings will be required to put in place resident engagement strategies which will allow residents to obtain information and be consulted about matters and decisions affecting the safety of their building (section 91).

Clause 156 of the Act makes amendments to the **Regulatory Reform (Fire Safety) Order 2005** that seek to strengthen fire safety requirements for non-domestic premises, which include the common parts of high-rise residential buildings subject to the new regime.

Further amendments will be made, including an increase in the level of fines available to the courts to the maximum level (unlimited), for the offences of impersonating a fire safety inspector and non-compliance with requirements imposed by an inspector and in relation to the installation of luminous tube signs.

MORE...

More information can be found at:

The Building Safety Act - GOV.UK (www.gov.uk)

Building safety - HSE

STUDY QUESTIONS

1. Describe the main means of enforcement action open to a fire authority.
2. Explain the following terms, as defined in the **RRFSO**:
   (a) Responsible person.
   (b) Relevant persons.
3. Outline the duties of the responsible person under the **RRFSO**.

(Suggested Answers are at the end.)

Identification of Hazards and the Assessment of Risk from Fire

IN THIS SECTION...

- Fire risk assessment assesses where and why a fire could start and what harm it would cause to those in and around the premises. Once this is established, measures can be put in place to prevent such fires.
- The five steps to fire risk assessment involve:
  - Identifying fire hazards to establish how a fire might start and what could burn.
  - Identifying people at risk and especially those vulnerable people at risk.
  - Evaluating the risk of fire breaking out and people being harmed as a result and removing or reducing the risk by the use of protective measures.
  - Recording the significant findings, implementing appropriate emergency plans, and providing information, instruction and training.
  - Regularly reviewing the assessment.

Fire Hazards and Assessment of Risk

Fire risk assessment can be defined as 'an organised and methodical look at premises, the activities carried on there, and the likelihood that a fire could start and cause harm to those in and around the premises'.

While there are no fixed rules about how a fire risk assessment should be conducted, it is important to adopt a structured approach that ensures all fire hazards and associated risks are taken into account.

The 'entry level' fire guidance document *A short guide to making your premises safe from fire* (Department for Communities and Local Government Publications, 2006), provides a good five-step framework for fire risk assessment. This approach is extended throughout a range of fire safety guidance documents, for different types of premises.

The Fire Risk Assessment Guides available are:

- *Offices and shops.*
- *Factories and warehouses.*
- *Sleeping accommodation.*
- *Residential care premises.*
- *Educational premises.*
- *Small and medium places of assembly.*
- *Large places of assembly.*
- *Theatres and cinemas and similar premises.*
- *Open-air events and venues.*
- *Healthcare premises.*
- *Transport premises and facilities.*

All fire hazards must be controlled

TOPIC FOCUS

"Risk" in relation to the occurrence of fire in a workplace depends on:

- The likelihood of a fire occurring.
- Its potential consequences for the safety of persons such as death, injury or ill health.

The types of physical harm that could be caused to persons by a workplace fire include:

- Smoke inhalation causing burning to the lungs and triggering conditions such as asthma.
- Suffocation or respiratory difficulties caused by depletion of oxygen.
- Poisoning by inhalation of toxic gases and other combustion products.
- Burning by heat, flames or explosion.
- Injury from falling or collapsing structures.
- Falls from a height or on the same level caused by panicking and rushing.
- Crushing injuries caused by panic or stampede.
- Injury from broken and flying glass.
- Mental or physical trauma.
- Death.

Five Steps to Fire Risk Assessment

Step 1: Identify Fire Hazards

For fire to occur, it is necessary to have three elements - fuel, heat (or ignition) and oxygen - come together at the same time, in the correct amount. This is known as the 'fire triangle', which we considered in Learning Outcome 11.3.

Identifying the fire hazards means looking for all the sources of ignition, fuel and oxygen that together might cause fire.

Step 2: Identify People at Risk

It is important to consider anyone who may be affected in the event of a fire, not just workers in the immediate area. Maintenance staff, contractors, other workers and people present outside normal working hours such as cleaners and security guards must also be taken into account.

Where visitors and members of the public, etc. have access to the premises, they must also be included in the assessment.

Additionally, any individuals or groups who may be particularly at risk must be considered. For example, young or inexperienced workers, people with mobility or sensory impairment, lone workers and pregnant workers, etc. may all be at greater risk in the event of a fire.

Step 3: Evaluate, Remove, Reduce and Protect from Risk

This is the process of evaluating:

- the risk of fire breaking out;
- how the resultant fire and smoke might travel through the premises; and
- the location of people relative to the fire and smoke.

The risk of fire breaking out can be removed or reduced by the use of appropriate fire prevention measures, such as:

- Effective control of ignition sources.
- Appropriate storage of flammable materials.
- Good housekeeping.
- Maintenance and inspection of equipment.
- Providing adequate information and training, etc.

The risk of people being harmed in the event that a fire does occur can be removed or reduced by the use of appropriate **fire precautions**, such as the provision of:

- Automatic detection and alarm systems.
- Adequate means of escape (including signage and emergency lighting).
- Fixed and portable fire-fighting equipment.
- Appropriate emergency procedures.
- Information and training, etc.

As it is rarely possible to entirely eliminate risk, it is important to ensure that remaining risks are controlled to an acceptable level.

STEP 1 Identify the fire hazards

STEP 2 Identify the people who might be at risk

STEP 3 Evaluate, identify and implement the fire precautions

STEP 4 Record findings, plan, instruct and train

STEP 5 Review and revise as necessary

Five steps to fire risk assessment

MORE...

A set of guides has been developed by the Department for Communities and Local Government to tell you what to do to comply with fire safety law, help you to carry out a fire risk assessment and identify the general fire precautions you need to have in place at:

www.gov.uk/workplace-fire-safety-your-responsibilities

The Publicly Available Specification (PAS) 79:2012 *Fire Risk Assessment - Guidance and a recommended methodology* provides a structured approach to fire risk assessment and gives a nine-step approach and corresponding documentation for carrying out and recording significant findings of fire risk assessments in buildings.

The Scottish Government has published Fire safety - existing non-residential premises: practical guidance. It covers fire safety responsibilities for business owners of non-residential premises andis available at:

https://www.gov.scot/publications/practical-fire-safety-guidance-existing-non-residential-premises-2/

Step 4: Record, Plan, Inform, Instruct and Train

The significant findings of the assessment, e.g. the fire hazards identified, individuals or groups at risk, the level of risk, and the actions taken to remove and/or reduce the risks must be **recorded**.

Appropriate **emergency plans** should be developed and implemented.

(Note that evacuation procedures should be practised regularly to ensure that all employees are familiar with the arrangements.)

Provide appropriate **information and instruction** to relevant persons on:

- Actions required to prevent fires.
- Actions required in the event of a fire.

Provide appropriate **training** for employees, in particular for those who may have specific duties in relation to fire prevention activities (e.g. conducting workplace inspections or checks on equipment, etc.) or in the event of a fire occurring (e.g. fire marshals).

Step 5: Review

The way we work is constantly changing - often as a result of new, or modifications to existing equipment, buildings, procedures, products and processes, etc.

As such, it is important to continue to be vigilant about new, or changed, fire hazards and risks and to review the fire risk assessment regularly.

How often is 'regularly' will depend on the extent of the risks and the degree of change but, as a guide, the assessment should be reviewed:

- Whenever there is reason to suspect it is no longer valid.
- After a significant or major incident.
- If there has been a significant change in circumstances in the workplace.
- Periodically, frequency depending on the nature of the business and the fire risks.

In all cases, the assessment should identify the hazards, assess the level of risk and consider how the risk may be minimised.

The risk assessment must be 'suitable and sufficient' and should be extensive and detailed enough to enable the responsible person to identify and prioritise the preventive and protective measures required to protect relevant persons from harm.

STUDY QUESTIONS

4. Describe the types of physical harm that could be caused to persons by a workplace fire.
5. Explain the method for carrying out a fire risk assessment.

(Suggested Answers are at the end.)

Fire Detection and Alarm Systems

IN THIS SECTION...

- All buildings should have provision for detecting fire and sounding an alarm in the event of fire.
- The type of detection/alarm system used will be determined by the type of occupancy and escape strategy.
- A fire detector identifies physical changes in the protected environment indicative of the development of a fire condition such as combustion products, visible smoke, flame/illumination or temperature rise.
- The types of detector designed to identify these conditions are:
  - Ionisation smoke detectors.
  - Optical detectors.
  - Radiation detectors.
  - Heat detectors.
- Manual alarm systems are suitable for small workplaces and include rotary gongs, hand strikers, handbells, whistles and air-horns.
- An automatic fire alarm system may be designed to respond to heat, smoke or the products of combustion and flames and may incorporate a facility for additional functions, such as closing down ventilation or air-conditioning plant, or activating automatic door releases.

Common Fire Detection and Alarm Systems and Procedures

Approved Document B to the **Building Regulations 2010** provides guidance on a general approach to fire safety in buildings and states that in order to determine the most suitable type of detection/alarm system for a particular building, it is necessary to determine the type of occupancy and escape strategy. The document recognises BS 5839-1, *Fire detection and fire alarm systems for buildings - Code of practice for design, installation, commissioning and maintenance of systems in non-domestic premises*, as a suitable source of guidance on the standard of automatic fire detection that may need to be provided within a building. BS 5839-1:2017 provides recommendations for the planning, design, installation, commissioning and maintenance of fire detection and fire alarm systems in and around non-domestic buildings. In addition, BS 5839-8:2013 gives corresponding recommendations for voice alarm systems which automatically broadcast speech or warning tones, in response to signals from their associated fire detection and fire alarm systems.

BS 9999:2017, *Fire safety in the design, management and use of buildings - Code of practice* also includes guidance on minimum levels of fire detection and fire alarm systems for premises. It acknowledges that premises with a higher fire growth rate may require a more sophisticated system, and in some low-risk premises an alternative means of giving warning in the event of fire might be more appropriate.

Where the escape strategy is based on simultaneous evacuation, activation of a manual call point or detector should cause all fire alarm sounders to operate. Where the escape strategy is based on phased evacuation, a staged alarm system might be more appropriate (i.e. one tone for alert, another for evacuate). This system is used extensively in large organisations, e.g. college and university campuses, where full evacuation is often not necessary.

Factors in the Design and Application of Fire Detection and Alarm Systems

All buildings should have provision for detecting fire. When a fire is detected, the following arrangements should be in place:

- Sounding an alarm in the event of fire.
- Evacuating staff to safe fire assembly points using means of escape routes.
- Fighting the fire.

Regular tests of an alarm system serve to check the circuits and to familiarise staff with the call note. The fire-warning signal must be distinct from other signals in use.

When selecting the method of alarm, it is important to ensure it will be heard in all parts of the building (the toilets, for example). Fitting fire doors to a building cuts down the distances over which call bells are heard, and may mean further bells must be fitted or noise levels raised.

A fire alarm can be raised automatically by a detection system or manually by a person in the affected building. We shall examine later the means available to somebody in the affected area for raising an alarm. These will generally be either manual or manual/electric, not forgetting that an alarm can always be raised by shouting.

Principal Components of Alarm Systems - Detection and Signalling

A fire detector identifies one or more physical changes in the protected environment indicative of the development of a fire condition. Usually mounted on ceilings or in air ducts, detectors are activated in the main by smoke or heat/light radiation. Such conditions can be readily identified:

- After ignition has occurred and the invisible products of combustion are released.
- When visible smoke is produced.
- When the fire produces flame and a degree of illumination.
- When the temperature in the vicinity of the fire rises rapidly or reaches a predetermined value.

The types of detector designed to operate at one of these particular stages are:

- Ionisation smoke detectors.
- Optical detectors.
- Radiation detectors.
- Heat detectors.

The most common types of detector system in use at present are those actuated by smoke and those actuated by heat.

The final choice is based on the risk to be protected against and the individual circumstances of each case (see the following table).

Automatic fire detectors

Type	Suitability
Smoke	
Ionisation Sensitive in the early stages of a fire when smoke particles are small. Sensitivity tends to drop as particles grow in size.	Areas with a controlled environment, i.e. free from airborne dust, etc., and generally housing complex equipment of a high intrinsic value, e.g. computer installations.
Optical Normally used as point detectors but have been developed to form zone sampling systems by monitoring air samples drawn through tubes.	Most effective in situations where the protected risk is likely to give rise to dense smoke (i.e. large particles).
Thermal Radiation	
Infrared Rapid detection because of almost instantaneous transmission of thermal radiation to the detector head. This is dependent, however, on the detector having a clear 'view' of all parts of the protected area.	Warehouses or storage areas, etc. Detectors are available which can scan large open areas and will respond only to the distinctive flame flicker. Can be used to detect certain chemical fires.
Ultraviolet As for infrared.	The ultraviolet detector tends to be used mainly for specialised purposes.
Heat	
Fusible alloys Alloys will need replacing each time detector operates.	Areas of general risk where vapour and particles are normally present. Cost is relatively low compared to other types of detectors.
Expansion of metal, air and liquid Generally self-resetting.	Both 'fixed-temperature' and 'rate-of-rise' are equally efficient but 'fixed-temperature' types are preferred in areas where a rapid rise in temperature is a likely result of the normal work processes.
Electrical effect	Not widely installed. Some specialist use.
Note for all types of heat detectors May be used as point or line detectors and are designed to operate at a pre-selected temperature ('fixed-temperature' type) or on a rapid rise in temperature ('rate-of-rise' type) or both. With all heat detectors (particularly 'fixed-temperature' types), 'thermal lag' needs to be considered when choosing the operating temperature.	'Rate-of-rise' types will compensate for gradual rises in ambient temperature and are more efficient than the 'fixed-temperature' type in low-temperature situations. ('Rate-of-rise' detectors generally incorporate a fixed-temperature device.)

Not all detectors will be equally sensitive in every possible situation. In some cases, a combination of different detectors may be required. Smoke and heat detectors are suitable for most buildings. Radiation detectors are particularly useful for high-roofed buildings, e.g. warehouses, and situations in which clean-burning flammable liquids are kept. Laser infrared beam detectors appear to have advantages where there are tall compartments or long cable tunnels, for example.

Such generalisations should be considered in conjunction with the nature of the risk to be protected against in order to establish the:

- Reliability required. A more robust detector is necessary in an industrial setting than is required for hotel purposes. Dusty or damp atmospheres will affect some detectors more than others.
- Sensitivity required. It would obviously be undesirable to install a smoke detector set at high sensitivity in a busy kitchen (or similar conditions).
- Location of detectors. The detectors should be located so they are in the best possible position to perform their function.

All alarm systems must be maintained and tested regularly, and the results recorded. Any faults discovered must be rectified and the system re-checked.

All staff must know how to raise the alarm and what to do when the fire alarm sounds.

Staff with hearing or other physical disabilities must be accommodated within an evacuation plan (e.g. people in wheelchairs cannot use stairs, as other people would, when a lift is inactivated). Emergency lights or vibrating devices may be used in addition to bells or sirens.

Manual and Automatic Systems

Manual Systems

Manual systems are suitable for small workplaces.

The purely manual means for raising an alarm involve the use of the following basic devices:

- Rotary gongs which are sounded by turning a handle around the rim of the gong.
- Hand strikers, e.g. iron triangles suspended from a wall accompanied by a metal bar which is used to strike the triangle.
- Handbells.
- Whistles.
- Air-horns.

These devices are normally found on the walls of corridors, entrance halls and staircase landings, where they are readily available to anyone who may need to raise an alarm. While they give an alarm over a limited area, operation of one of them is rarely adequate to give a general alarm throughout the premises. As a person is required to operate them, a continuous alarm cannot be guaranteed for as long as may be necessary.

Manual/Electric Systems

These are systems which, although set in motion manually, operate as part of an electrical alarm circuit. The call points in a manual/electric system are invariably small, wall-mounted boxes which are designed to operate either:

- automatically, when the glass front is broken; or
- when the glass front is broken and the button pressed in.

Most available models are designed to operate immediately when the glass front is broken.

Fire alarm

In order to raise an alarm, it is possible to use facilities which may already be installed in a building for other purposes, e.g. a telephone or public address system. With automatic telephone systems, arrangements can be made for a particular dialling code to be reserved for reporting a fire to a person responsible for calling the fire and rescue service and sounding the general alarm. Alternatively, it can be arranged that use of the code automatically sounds the general alarm.

Automatic Systems

An automatic fire alarm system may be designed to respond to heat, smoke or the products of combustion and flames. Although the system will give warning of a fire, it cannot take action to contain it. However, some more elaborate designs do incorporate a facility for additional functions, such as closing down ventilation or air-conditioning plant, or activating automatic door releases.

Reliability and False Alarms

We have already noted that all buildings should have provision for detecting a fire and sounding an alarm to initiate the evacuation of people to a place of safety. For this arrangement to be effective, the alarm system must be reliable and only activate in a fire situation. Frequent false alarms will only serve to undermine the system and lead to occupants disregarding any alarm and failing to evacuate the premises.

If the detector initiates some control action (such as plant shut-down or sprinklers), to reduce the potential for false alarms, a **voting system** can be incorporated (i.e. several detectors are required to set off the alarm and subsequent control action), but this will require a higher density of detectors. Typically, voting systems may be set to trigger if two out of three detectors are activated, but it depends on the perceived level of risk. In some cases, it can be two out of 16 or three out of 75.

False alarms can arise from unwanted activation of the detector caused by:

- Wrong choice of head:
  - Smoke detector set at high sensitivity in a busy kitchen.
- Wrong positioning of detector:
  - Thermal radiation detectors out of sight of the protected area.
  - A single ultraviolet flame detector in the vicinity of an arc welding area.
- Lack of control of working environment:
  - Smoking.
  - Hot work.
  - Dust from spillages or maintenance.
- Malicious actuation:
  - Malicious operation of 'break glass' call points.
  - The intentional directing of smoke into smoke detectors.
- Wiring defects:
  - Poor installation.
  - Lack of maintenance.
- Deterioration of the system:
  - Corrosion.
  - Damp environments.

STUDY QUESTIONS

6. Name the four types of detector commonly used in buildings.
7. What are the benefits of regular testing of fire alarms?
8. For what types of workplaces are manual alarm systems suitable?

(Suggested Answers are at the end.)

Fixed and Portable Fire-Fighting Equipment

IN THIS SECTION...

- Fixed fire-fighting systems fall into four main categories:
  - Sprinklers.
  - Drenchers.
  - Total flood systems.
  - Deluge systems.
- The main types of extinguishing agent used in fixed installations are:
  - Water.
  - Foam.
  - Carbon dioxide.
  - Halon.
  - Dry powder.
  - Wet chemical.
- Fires are classified into different categories - Class A, B, C, D and F - to help in the selection of extinguishing agent.
- Extinguishing a fire is based on removing one or more sides of the fire triangle:
  - Removing the fuel by starvation.
  - Removing the oxygen by smothering.
  - Removing the heat by cooling.
- Fire-fighting equipment should be sited in an easily seen and reached position, regularly inspected and maintained, and persons required to use it suitably trained.
- Environmental damage can result from contaminated fire-fighting water run-off - sites should consider the polluting effects of fire in their emergency plans.

Design and Application of Fixed Fire-Fighting Systems and Equipment

In certain establishments, fixed fire-fighting systems are installed. Fixed (sometimes referred to as 'passive') fire-fighting systems fall into four main categories. Those systems dedicated to the protection of life have a higher specification than those dedicated to property.

Sprinkler

The four categories are as follows:

- **Sprinklers** are characterised by independent, sealed sprinkler heads. Only the discharge heads in the immediate vicinity of the fire rupture, so that water damage is limited. They are widely used and have a rating ranging from Light Hazard (LH) through Ordinary Hazard (OH) to High Hazard (HH).

- **Drenchers** are designed to protect adjacent buildings or facilities from the effects of radiated heat and burning embers from a fire. They provide a curtain of water over parts of a building (or openings). They are commonly used to protect large gas storage tanks.
- **Total flood systems** render the atmosphere inert by dilution or flame interference (CO_2 or halon). They are dangerous to occupants and therefore a safe system (manual lock-off) must be employed. There is also a danger of re-ignition when opened up.
- **Deluge systems** have all the discharge heads open but the flow of extinguishing agent is controlled by a single deluge valve (which may be activated by a pilot sprinkler system). They are often used for high-risk cases such as flammable liquid storage tanks.

Fixed fire-fighting systems consist primarily of pipework which delivers and releases an extinguishing medium when activated directly by heat or indirectly by the warning/alarm system. Such systems are likely to be installed in large buildings where there is a high risk, where access is difficult, or where equipment or stock is valuable.

The main types of extinguishing agent used in fixed installations are:

- **Water**, which is generally used in sprinkler systems.
- **Foam**, which includes either:
  - low expansion foam, which is suitable for flammable liquid fires; or
  - high expansion foam, which is especially useful in inaccessible areas, e.g. cable tunnels and basements.
- **Carbon dioxide**, which is suitable for hazardous plant, e.g. electrical equipment, computer areas, control rooms and sensitive materials.
- **Halon**, which is used in similar situations to carbon dioxide. There has been movement away from the use of halon over recent years due to its potential effect on the ozone layer and other undesirable environmental effects; it is now little used because its use is strictly controlled (limited to specialist military and aerospace uses). (Alternatives to the use of halon include the use of water mist.)
- **Dry powder**, which is suitable for flammable liquids, electrical equipment, or situations where water damage must be kept to a minimum (dry powder is not suitable where re-ignition may occur).
- **Wet chemical**, which is used for high-temperature fat fires (Class F) and so its use is restricted in fixed installations to those that protect equipment using high-temperature fats and oils.

For water-based systems, a supply of water is clearly required. There are two basic systems:

- **Wet riser** systems remain filled with water at fire mains pressure. Such systems are subject to frost damage in unprotected areas, corrosion and water leaks.
- With **dry risers**, water is only available at the outlets when the system is connected either to the fire mains or from a pumped reservoir or supply. The system is not subject to frost damage - and so may be used in cold stores.

Wet and dry riser sprinkler systems have independent, sealed sprinkler heads. The sprinkler heads are designed to open under fire conditions (e.g. through fusible solder links). Water is only, of course, directed to those heads which have been activated - since they are all independent.

Deluge systems are designed for high-risk areas and consist of pipework with open heads, the water being held back by a deluge valve. A fire-detection device trips the deluge valve and water is discharged from every head.

Classification of Fires

Fire classification is a system of categorising fires taking into account the type of material and fuel available for combustion.

Class letters are often assigned to the different types of fire but globally, there are many separate standards for the classification of fires (e.g. Indonesia has a class E - electrical class). The UK classification system is illustrated in the table below but care has to be taken when working in other countries to establish the classification system in that region.

Nonetheless, the classification system serves the same purpose overall in determining the type of extinguishing agent that can be used for the fuel involved in the fire.

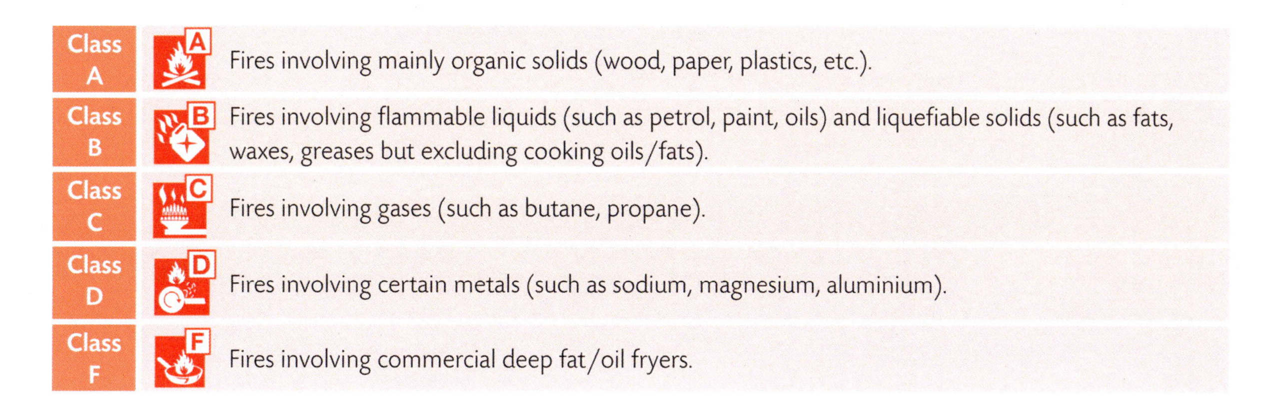

Class A	Fires involving mainly organic solids (wood, paper, plastics, etc.).
Class B	Fires involving flammable liquids (such as petrol, paint, oils) and liquefiable solids (such as fats, waxes, greases but excluding cooking oils/fats).
Class C	Fires involving gases (such as butane, propane).
Class D	Fires involving certain metals (such as sodium, magnesium, aluminium).
Class F	Fires involving commercial deep fat/oil fryers.

Portable Fire-Fighting Equipment

The range of fire extinguishers (their size, colour, method of operation and claims for performance) is so great that it can be confusing. The equipment which the average person will use in the event of fire should be **suitably located** and **suitable for the risk**. The problems arise when more than one type of risk may be encountered and the person, who is operating under pressure, is faced with a choice of extinguisher. It could well be that the wrong choice could render the efforts wasted or even expose the person to danger!

Firstly we should identify the **nature of the risk** and the choices of agent which are available.

The nature of risk

Fire Class	Description		Examples	Extinguishing Agents
A	**Solid materials** Usually of organic origin (containing carbon-based compounds)		Wood, paper, fibres, rubber	Water; foam; dry powder (ABC)
B	**Flammable liquids and liquefied solids**	Those miscible with water (capable of being mixed)	Alcohol, acetone, methyl acetate	Dry powder; specialist foam; CO_2
		Those immiscible with water	Petrol, diesel, oil, fats and waxes	Dry powder; foam
C	**Gases and liquefied gases**		Natural gas, liquefied petroleum gases (butane, propane)	Turn off the supply; liquid spills may be controlled by dry powder
D	**Flammable metals**		Potassium, sodium, magnesium, titanium	Special powders (m28 or l2); dry sand or earth; graphite powder; sodium carbonate and salt and/or talc
F	**High temperature cooking oils**		-	Specialist 'wet chemical'; fire blanket (minor fire only)

Gas fires can be difficult to deal with. While dry powder and carbon dioxide may be used to knock the flame down, there is a risk of a build-up of gas if it cannot be turned off. In some situations, it may be preferable to allow the fire to continue and to call the fire and rescue service.

You will notice that electrical fires are not listed; this is because electricity is not a fuel, it will not burn. However, it can cause fires and it can be present in fires so we have to consider it when fighting fires.

Extinguishing Media and Mode of Action

Extinguishing a fire is based on removing one or more sides of the fire triangle:

- **Removing the Fuel**

  Extinction by this process is known as '**starvation**'. This can be achieved by taking the fuel away from the fire, taking the fire away from the fuel and/or reducing the quantity or bulk of fuel available. Materials may therefore be moved away from the fire (to a distance sufficient to ensure that they will not be ignited by any continuing radiant heat) or a gas supply may be turned off.

- **Removing the Oxygen**

  Extinction by this process is known as '**smothering**'. This can be achieved by either allowing the fire to consume all the available oxygen, while preventing the inward flow of any more oxygen, or adding an inert gas to the mixture. The most usual method of smothering is by use of a blanket of foam or a fire blanket.

Standard, red-bodied extinguisher

- **Removing the Heat**

  Extinction by this process is known as '**cooling**'. Cooling with water is the most common means of fighting a fire and this has a dual effect:

  - Absorbing the heat and thereby reducing the heat input into the fire.
  - Reducing the oxygen input, through the blanketing effect of the steam produced.

Although water is the most common medium used to fight fires, it is by no means either the only or the most suitable substance. Indeed, using water on certain types of fire can make the situation worse.

The main different types of extinguishing media are described below and you should note their application to the classification of different types of fire.

Water

Water, applied as a pressurised jet or a spray, is the most effective means of extinguishing class A fires, and may also be used, as a spray, on class B fires involving liquids and liquefied solids which are miscible (capable of mixing) with water, such as methanol, acetone and acetic acid. While ineffective on class C fires themselves (those involving gases), water may be used to cool leaking containers.

It must **never** be used on:

- Fires involving electricity, as the current can flow up the stream of water.
- Non-miscible (fat or oil) liquid fires, as only a cupful of water can cause the whole fire to erupt into a conflagration.

Foam

Foam is a special mixture that forms a smothering blanket over the fire, cutting off the supply of oxygen. It can be used on class A and B fires (although there are some restrictions in its use on class B fires since certain types of foam break down on contact with alcohols) and also on small liquefied gas fires (which make up certain class C fires).

Using foam as an extinguishing agent demands considerable skill when dealing with anything but very small-scale liquid fires, since the procedure is to start at the rear and to lay a blanket of foam across the surface of the liquid.

Dry Chemical Powder

Usually ABC Dry Powder - so called because it is suitable for use on class A, B and C fires (although it is rarely the best extinguishant for class A).

The powder is sprayed as a cloud over the fire, again acting to smother the supply of oxygen. It can be used on class B fires and on small liquefied gas fires (within class C).

Specialised dry powders using inert substances are also used on class D fires, where they form a crust over the burning metal and exclude the oxygen.

Dry powders are also effective on fires involving electricity.

Carbon Dioxide Gas

This again works by means of smothering the supply of oxygen. It is effective on class B fires and also for electrical fires as the gas can enter into the inside of the equipment.

It should not be used on Class D fires due to the danger of the hot metal exploding.

Wet Chemical

This is used for high-temperature fat and oil fires (Class F). The liquid chemical undergoes saponification by the action of heat in contact with the high-temperature fat, causing a surface layer of soapy foam to form that prevents the escape of vapour from the surface of the liquid.

Vaporising Liquids

When applied to a fire, these agents produce a heavy vapour which extinguishes the fire by excluding oxygen. They are safe to use on class A and B fires, and are particularly effective on fires involving live electrical equipment since they interfere with electrical combustion reactions.

Halon 1211 is a vaporising liquid which used to be widely used in portable fire-fighting equipment. Halons have been banned in many countries since 1994 and are being phased out in others.

Identification of Fire Extinguishers

In order to identify the different types of fire extinguisher in common use, they are colour-coded depending on the type of extinguishant inside. There are currently two systems in use:

- Older fire extinguishers will have the whole body colour-coded.
- Newer fire extinguishers will have a red body with a coloured band or label.

The following table shows the colour coding for the different types of extinguishant:

Colour coding for fire extinguishants

Fire Extinguisher Content	Colour of Body or Label/Band
Water	Red
Foam	Cream
Dry Powder (ABC)	Blue
Dry Powder (D)	Violet
Carbon Dioxide	Black
Wet Chemical	Yellow

Siting of Extinguishers

The correct type of extinguisher should be available for the risk it is going to protect against. The fire-fighting equipment should be sited in an easily seen and reached position, usually by an escape route. The location should be marked and should not be further than 30m from an alternative equipment location (see the following figure).

Siting of extinguishers

The extinguishers should be located so they are:

- Conspicuous.
- Readily visible on escape routes.
- Properly mounted.
- Accessible (less than 30m from any fire).
- Sited to avoid temperatures beyond the operating range and corrosive environments. Special extinguishers should be sited close (but not too close) to the risk.

BS 5306-8:2012, *Fire extinguishing installations and equipment on premises - Selection and positioning of portable fire extinguishers - Code of practice*, gives guidance on the suitability and positioning of portable fire extinguishers.

Inspection and Maintenance

The main relevant guidance covering this subject is BS 5306-3:2017, *Fire extinguishing installations and equipment on premises - Commissioning and maintenance of portable fire extinguishers - Code of practice.*

Regular (at least monthly) **visual inspections** should be carried out. This is a straightforward check that the extinguisher:

- is still there,
- is unused and undamaged,
- has not been tampered with,
- is accessible and visible,
- has instructions on it that are legible; and
- that any pressure gauges indicate that it is within operational limits.

In addition, all portable fire extinguishers will need to undergo an annual basic service and also:

- either an overhaul every ten years (carbon dioxide types only); or
- an extended service (all other types - typically this is required every five years).

An extended service includes a test discharge of the extinguisher. An overhaul includes a test discharge and also a pressure test of the body shell.

We will look at the requirements for the basic service. These vary depending on the mode of operation (e.g. stored pressure vs. cartridge) and/or the extinguishing agent.

Stored Pressure Type

These are pressurised containers. The extinguishing agent is expelled when the pressure is released using the operating trigger. Water, water-based (e.g. foam and wet chemical types), powder and primary sealed powder are commonly found as stored pressure types.

Primary sealed powder extinguishers are a type of stored pressure extinguisher where the head can be detached without releasing the pressurised powder (because it has an additional seal which is only pierced during activation of the extinguisher). The basic maintenance regime is the same for all these types of stored pressure devices and comprises the following checks (summarised from Table D.2, Annex D of BS 5306-3), which should be carried out in order:

- Check the safety clip and indicators to see if the extinguisher has been used.
- Check that the pressure is correct (many extinguishers have a pressure gauge fitted, otherwise a pressure gauge may be attached to a special connection).
- Check for damage (corrosion, dents, etc.) and replace as necessary.
- Weigh the extinguisher (losses of more than 10% need a full recharge).
- Check the operating instructions (legible, correct).
- Where the operating head is designed to be removable without discharging the contents (e.g. primary sealed powder type), do so and check/clean/lubricate the operating mechanism.
- Remove and replace the safety pin, checking operating lever for damage and free movement.
- Check hoses, discharge horn and nozzle are not blocked and there is no damage. Replace the seals for these components.

The extinguisher is reassembled on completion of the checks and details are completed on the maintenance label. An inspection report will detail the current state of the extinguisher.

Carbon Dioxide

The basic service regime should be similar to the above with the exception of the second point in the list (pressure is not checked for carbon dioxide extinguishers).

Because they are classified as 'pressure vessels', carbon dioxide (CO_2) extinguishers need an extended service every 10 years. This involves a hydrostatic pressure test, to comply with pressure vessel legislation and a new valve. Most fire extinguisher servicing engineers will supply a 'service exchange' (refurbished) unit due to the onerous requirements of this process.

Cartridge Operated

Rather than the propellant being in the same chamber as the extinguishing agent, these types of extinguishers are fitted with a separate cartridge which contains the propellant. Piercing the cartridge releases the propellant which, in turn, expels the extinguishing agent. Water, water-based and powder are commonly found as cartridge types (as well as stored pressure types).

The basic service should be the same as for the stored pressure type (as outlined above) **except** the pressure check is not relevant (so not carried out) **and** the following additional checks are made:

- Open the extinguisher (unscrew the head), removing the gas cartridge, and then:
  - For water/water-based extinguishers, pour out the contents from the body and check interior for corrosion (or, if fitted, remove and check any inner container for leakage). The original charge may be re-used after inspection if the liquid is still in good condition, otherwise replace with fresh.
  - For powder extinguishers, check the powder for caking, lumps, etc. and re-charge if necessary. Check the interior body for corrosion, etc.
- Check operating mechanism (for free movement) and air passages (should be unobstructed).

Note: most fire extinguishers need an extended service every five years. This means completely discharging the extinguisher, checking for internal corrosion, refilling and repressurising.

Training Requirements

The ability to carry out fire-fighting with portable extinguishers may not only control the rate at which a fire spreads, thereby giving those precious few moments which mean the difference between a person escaping or becoming a victim, but often reduces fire damage to a lower level than would have been the case if the fire had proceeded without being checked. It is therefore very important that all personnel are familiar with the fire-fighting equipment and are able to use it correctly.

The following points form a general scheme for training in the use of fire-fighting equipment:

- General understanding of how extinguishers operate.
- The importance of using the correct extinguisher for different classes of fire (ideally, only the correct type of extinguisher should be available for use in any particular situation).
- Recognition of whether the extinguisher has to be used in the upright position or in the upside-down position.
- Practice in the use of different extinguishers. This can be done with or without a practice fire; dealing with a live fire is obviously the better method.

- Personnel must understand that they can only provide a 'first-aid' treatment and evacuating the building must take precedence over fighting a fire if the condition demands immediate evacuation.
- When, and when not, to tackle a fire. If the fire is small and has not involved the building structure, then portable extinguishers can generally be used. Always see that a means of escape is maintained.
- When to leave a fire that has not been extinguished. As a general rule, once two extinguishers have been discharged, the fire requires the fire and rescue service.
- When leaving an unextinguished fire, try to close all doors and windows to help contain the fire.

Environmental Considerations

Environmental damage can result from contaminated fire-fighting water run-off:

- Fire-fighting water run-off can carry large amounts of chemicals.
- Perfluorooctane sulfonate (PFOS), is a chemical used in some fire-fighting foams which does not break down in the environment. It accumulates in organisms and works its way up the food chain.
- PFOS accumulates heavily in humans and animals, and in humans has a relatively long half-life in the body. It has been found to be bioaccumulative and toxic to mammals.
- Pollutants including PFOS seeped into the water table and two local rivers were found to have been contaminated.
- Run-off can make its way into drains, rivers and sewage treatment works.
- Groundwater contamination may take several months to become apparent, and may persist for many years.

Run-off containing large amounts of chemicals

Fire-fighters used more than 250,000L of foam to contain the Buncefield fire in December 2005, which destroyed 20 huge oil storage tanks and created a black cloud 200 miles wide.

MORE...

The European Chemicals Agency (ECHA) has produced a report into the use of PFAS and flourine free alternatives in fire-fighting foams.

See: https://echa.europa.eu/documents/10162/28801697/pfas_flourine-free_alternatives_fire_fighting_en.pdf/d5b24e2a-d027-0168-cdd8-f723c675fa98

Legal Obligations Related to Environmental Protection in the Event of a Fire

- The **Environmental Permitting (England and Wales) Regulations 2016** cover discharges to surface water in England and Wales. The Regulations state that:

  "A person must not, except under and to the extent authorised by an environmental permit... cause or knowingly permit a water discharge activity or groundwater activity."

  It is likely that, if an organisation pollutes a river or other surface watercourse, it will be prosecuted under these Regulations.

- The duties of the **Environment Agency (EA)** (**Natural Resources Wales** in Wales) include prosecutions for polluting "**controlled waters**" which include:
  - **Territorial waters** extending to three nautical miles.
  - **Coastal waters**.
  - **Inland freshwaters**: rivers, streams, lakes, etc.
  - **Groundwater**.
- The maximum penalties which can be imposed:
  - Magistrates' Court:
    - Unlimited fine and/or six months' imprisonment.
  - Crown Court:
    - Unlimited fine and/or five years' imprisonment.
- The **EA** has an operating agreement with the **fire service** to ensure a co-ordinated effort in the event of fire incidents that have potential to pollute controlled waters and the disposal of attendant waste such as contained, contaminated fire-fighting water run-off.

  This general agreement means that:
  - The **EA** takes responsibility for:
    - **Remedial action** in cases of water pollution.
    - **Regulation of any waste issues** arising.
  - The **Fire Service** takes responsibility for:
    - **Preventing the environmental impact** of the incident, so far as reasonably practicable, but recognising that there may be higher priorities that may prevent this, such as the protection of life.
    - **Informing the EA of water pollution incidents** or where this could potentially happen.

  The EA has in the past also provided front-line pollution control equipment (such as drain blockers, booms, sealing putty, absorbents) and training to the fire service.

Pre-Planning the Minimisation of Environmental Impact of Fire and Containment Procedures

Sites should consider the polluting effects of fire in their emergency plans. In case of a fire, there is the possibility of escape of contaminated fire water:

- Directly to surface run-off, into rivers and the ground.
- Via the site's surface water drainage system.
- Via the foul sewage system and out unaltered through the treatment works.

River polluted by fire-fighting water run-off

To mitigate the effects of the incident, pre-planning of containment is required which depends on the following:

- The **hazardous nature** of substances on site.
- The **risk of fire**.
- Sensitivity of the **receiving environment**.

- The **reasonable practicability of the solution** such as dimensions of the site and the cost involved.
- **'Built-in' permanent remote containment systems** for a site include:
  - **Lagoons** - earth-banked containment basins:
    - Effective at containing firewater run-off, provided they are impermeable and incorporate some sort of isolation from the drainage system.
  - **Purpose-built tanks** - to intercept run-off:
    - Can be placed below or above ground.
    - Include as a last resort the possibility of using storm tanks at the sewage treatment works.
  - **Shut-off valves** - used to isolate part of a site's drainage system.
- **Emergency remote containment systems** (when permanent solutions are not practicable) include:
  - **Sacrificial areas** - run-off is conveyed to a designated 'sacrificial' area:
    - Involves use of permeable soil or porous media with an impermeable lining.
  - **Temporary bunding** of impermeable car parks - e.g. using sandbags.
  - **Portable tanks and tankers**:
    - Requires temporary blocking of the site drainage system to set up a temporary sump from which the run-off can be pumped into the tank.
    - Also allows the possibility of re-use of the run-off as fire-fighting water.
- **Site and damaged area clean-up considerations** - having contained the fire-fighting water run-off, the last stage in minimising the environmental impact of fire and fire-fighting operations is to ensure that the area is cleaned up satisfactorily. This will principally rely on ensuring that collected contaminated run-off is disposed of properly, in accordance with prevailing waste regulations.

STUDY QUESTIONS

9. Describe the five basic classes of fire.
10. Identify the colour coding of the five common types of fire extinguishers and the types of fire that they are suitable for.
11. With reference to fire extinction, describe what is meant by 'starvation'.
12. Explain why gas fires may be difficult to deal with.

(Suggested Answers are at the end.)

Means of Escape

IN THIS SECTION...

- The main factors to consider in provision and maintenance of means of escape are the:
  - Nature of the occupants.
  - Number of people attempting to escape.
  - Distance they may have to travel to reach a place of safety.
  - Size and extent of the 'place of safety'.
- Travel distance is related to the mobility of the occupants and the danger of fire spread. If the travel distance exceeds 12m, two alternative escape routes should be provided.
- All stairways that comprise part of the means of escape must be constructed and arranged to permit persons using the route in an emergency to pass freely.
- Escape routes must take into account predicted fire, heat and smoke spread that will follow as a fire develops.
- Fire doors are incorporated for smoke control and/or to protect means of escape or to segregate areas of special risk.
- Emergency (or safety) lighting should be provided where failure of the normal system would cause problems in using the means of escape. The purposes of emergency lighting are to:
  - Identify the escape route.
  - Provide illumination along the route and the exits.
  - Ensure that alarm call points and fire-fighting equipment can be easily located.
- Landlords and owners of multi-occupancy buildings have a responsibility under the **RRFSO** to maintain fire safety in communal areas.

Provision and Maintenance of Means of Escape

To **protect the occupants**, a building must incorporate features which will protect them **while they escape from a fire**. Likewise, the building must not incorporate features which would endanger the occupants or prevent their escape in the event of a fire. The main factors to consider in provision and maintenance of means of escape are the:

- Nature of the occupants, e.g. mobility.
- Number of people attempting to escape.
- Distance they may have to travel to reach a place of safety.
- Size and extent of the 'place of safety'.

Means of escape

The total of these considerations results in an **adequate means of escape**, by which, under the worst conditions, the exposure of the occupants to danger will be minimal.

General Requirements

Travel Distances

This is the distance from an occupied area of the building to an exit leading to a place where a person would no longer be in immediate danger. This place can be a place of **relative safety** such as a protected stairway or corridor, or may be a place of safety outside the building.

The travel distance can be related to the mobility of the occupants and the danger of fire spread. The occupants should not be exposed to fire danger for longer than two to three minutes in normal office-type conditions. Under favourable conditions, the assumed speed of travel, for mobile persons, is 12m/min.; this may be increased for ground floors to 18m/min. If there is a problem in meeting the time or rate of travel, or if there is a particular fire-spread danger, the distance will be reduced.

The factors which have to be considered when assessing **means of escape** will vary widely from one set of premises to another. The figures shown in the following table are only guidelines, as individual circumstances vary considerably.

Maximum horizontal travel distances
Source: Building Regulations Approved Document B

Use of the premises or part of the premises		Maximum travel distance where travel is possible in:	
		One direction only (m)	More than one direction (m)
Institutional		9	18
Other residential:			
in bedrooms		9	18
in bedroom corridors		9	35
elsewhere		18	35
Office		18	45
Shop and commercial		18	45
Assembly and recreation:			
buildings primarily for disabled people		9	18
areas with seating in rows		15	32
elsewhere		18	45
Industrial	Normal Hazard	25	45
	Higher Hazard	12	25
Storage and other non-residential	Normal Hazard	25	45
	Higher Hazard	12	25

BS 9999:2017 *Fire safety in the design, management and use of buildings - Code of practice* includes specific guidance on travel distances based on risk profile and fire protection measures provided.

Remember that the distance will not necessarily take people out of the building, but to a **place of relative safety**, placing a fire door between them and the fire. Once within this place, the structure will protect them from smoke and heat while they make their way to a place clear of risk.

If the travel distance exceeds 12m, two alternative escape routes should be more than 45° apart from any point. Some examples are shown in the following figure:

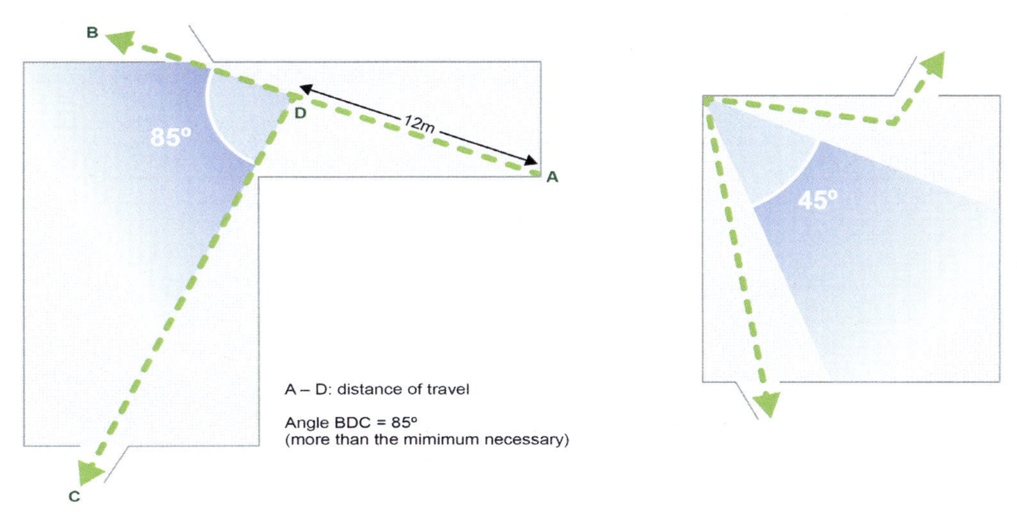

Distance of travel more than 12m

Where a building is divided into a number of rooms with linking corridors, it will be necessary to consider **means of escape** in each of the following ways:

- Travel within rooms.
- Travel from rooms to a storey exit.

Distance of travel should be measured as being the actual distance to be travelled between any point in a building and the nearest **storey exit**.

An example of **distance of travel** in a factory with both high and low fire-risk areas is shown in the following figure:

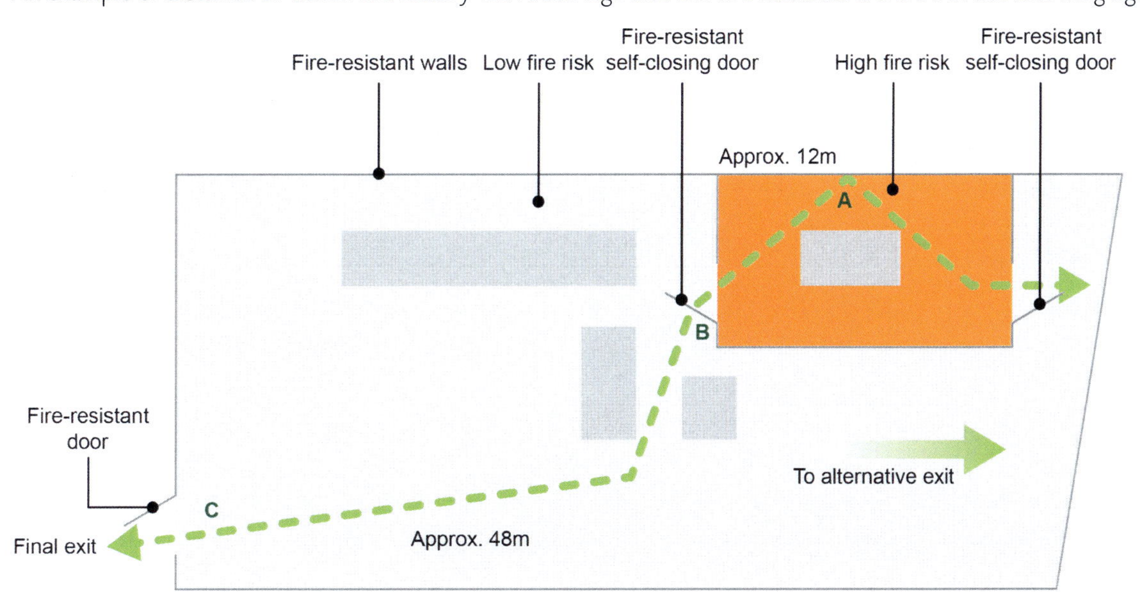

Distance of travel in a factory

Stairs

All stairways that comprise part of the means of escape must be constructed and arranged to permit persons using the route in an emergency to pass freely. The following are the standards generally accepted for a well-designed staircase:

- Sited on an outer wall with good lighting and ventilation.
- Constructed to discharge to open air or into a protected lobby which discharges direct to open air.
- Fully enclosed with the minimum fire-resistive material required.
- Sufficient width for escaping occupants.
- Handrails.
- No more than 16 and no less than three steps, without a landing.
- The pitch should comply with **Building Regulations** requirements.
- No more than 36 steps and landing without a turn.
- Clear headroom, no projection below two metres.
- Spaces below flights should be open, or closed without access.
- Non-combustible or class A linings.
- Non-slip/trip surfaces.

Ideally, stairways should be at least 1.05m wide; and where wheelchair users may be present, exits should be not less than 900mm wide.

In general, all means of escape should have **clear headroom of not less than 2m**. There should be no projections from any wall or ceiling below this height (except for door frames) which could impede the free flow of persons using the escape route.

A **continuous handrail** should be fixed on **each side** of all stairways, steps, landings and ramps at a height suitable for the average person, e.g. about 840mm to one metre above the pitch line of the nosing of steps.

In certain circumstances, escape stairs may require a smoke control system (typically this would involve either pressurisation or ventilation). For example, a stairway which is the only one serving a building which has more than one level above or below ground would ordinarily also require a protected lobby; a smoke control system (designed in accordance with BS EN 12101-6:2005) is an allowable alternative.

Passageways and Doors

Escape Corridors

- **Partitions**

  Any corridor which forms part of a means of escape route should be kept free of fire and smoke. The lining of any escape corridor should be non-combustible or class A.

- **Glazing**

  It is usual practice to leave a space of about 1.05m between the bottom of the glazing panel and floor level, to enable people who have to pass the glazing to crawl along below the level of radiated heat.

- **Variation in Width**

  A corridor should be wide enough to allow all persons to escape safely. It must not get narrower. Any narrowing or bottleneck conditions can seriously affect evacuation and may lead to panic conditions. The width of the corridor, in route calculations, is its narrowest point.

- **Surfaces**

  All floor surfaces to route corridors should be of non-slip material. There should be no tripping/slipping hazards caused from steps, slopes or mat wells.

Fire Exits and Doors

As well as having a sufficient number of exits, the design of routes also needs to take into account the predicted fire, heat and smoke spread that will follow as a fire develops.

The main effects to consider in the movement of smoke and heated gases are:

- **Horizontally** - relatively slowly, with considerable intermixing and billowing.
- **Vertically** - rapidly in convection currents.

The principles of smoke-stopping are based on these two forms of smoke spread.

Fire doors can be primarily for smoke control and/or to protect means of escape or to segregate areas of special risk.

A **fire-resisting**, **self-closing door** which fits tightly in its frame will stop smoke and superheated gas travel within a building. The door frame must have rebates which are sound and prevent heat and smoke creeping around the tight fit between it and the door.

Circulation areas (corridors, staircases) will form the main escape routes, so it must be ensured that fire cannot occur in, or spread into, these areas. This is achieved by **all linings in these areas being non-combustible** (class 0 **Building Regulations**). Furnishings, drapes, display units and all forms of storage should be discouraged.

Fire-resisting doors (smoke-stop) are used to:

- Break corridors into sections and thereby reduce the area of smoke logging.
- Separate stairs from the remainder of the floor area.
- Confine an outbreak of fire to its place of origin.
- Keep escape routes free of smoke for long enough to permit evacuation of occupants.

Construction of Doors

Any fire door (and the partition in which it is hung) should be rated to the fire-resistive standards required by the fire authority/building control. The workmanship must be to a high standard. Any glass panels within a fire door should also be fire-resisting and limited in size.

A well designed timber fire door will delay the spread of fire and smoke without causing too much hindrance to the movement of people and goods.

Every fire door is therefore required to act as a barrier to the passage of smoke and/or fire to varying degrees, depending on its location in a building and the fire hazards associated with that building.

The main categories of fire doors are FD30 and FD60 fire doors which offer 30 and 60 minutes' fire protection. To determine the FD rating of fire doors, the manufacturers have their fire doors assessed by subjecting them to a test procedure as specified in BS 476-22:1987 or BS EN 1634-1:2014.

All fire doors must be fitted with the appropriate intumescent seals designed to expand under heat and fill the gaps between the door leaf and frame, thereby preventing the passage of smoke and fire to other parts or compartments of the building.

With false ceilings above fire-resistive partitions that hold fire doors, the partition should extend through the void and be sealed to the underside of the floor above.

The self-closing device must be positive and durable. The door must close with a close fit to the frame.

Opening of Doors

- Means of escape doors should normally open **in the direction of exit**.
- A door which opens directly over steps or into a corridor should be **recessed to at least the width of the door**. This will prevent it obstructing the corridor or stairs and injuring passers-by.
- **Revolving doors** are not permitted on escape routes unless a clearly indicated pass door is provided beside them.
- **Rolling shutters**, when closed, can form an obstruction to means of escape when people are in the building. They should either be locked in the open position, or be provided with a clearly indicated pass door beside them.
- **Sliding doors** may be permitted with certain restrictions. They should not be positioned at the foot of stairs and never be used in assembly occupancies. The door should be clearly marked 'Slide to open', with an arrow showing the direction to open.
- **Pass doors** between different parts of the same building can be useful, providing satisfactory means of escape is available from both sides.
- **Intercommunicating doors and wall hatches** provide a good means of escape at low cost. They should be clearly marked: 'Fire Exit - Do Not Obstruct', on both sides and lack of obstruction should be ensured.

Door Fastening

All fire doors, except those to cupboards and service ducts, should be fitted with effective self-closing devices to ensure positive closure of the door. Rising butt hinges are not normally acceptable. Fire doors to cupboards, service ducts and any vertical shafts linking floors should either be self-closing or be kept locked shut when not in use and labelled accordingly.

In some buildings, self-closing fire doors may cause difficulties for staff and members of the public. **Automatic door releases** can be used to hold open such doors provided that:

- The door release mechanism fails to safety (i.e. in the event of a fault or loss of power, the release mechanism is triggered automatically).
- All doors fitted with automatic door releases are linked to an automatic fire-warning system appropriate to the fire risk in the premises.
- All releases are automatically triggered by any one of the following:
  - The actuation of any automatic fire detector.
  - The actuation of any manual fire-alarm call point.
  - Any fault which renders the fire-warning system inoperable.
  - The isolation of the alarm system for any reason, e.g. maintenance.
- Doors so fitted are capable of being closed manually.
- The release mechanisms are tested at least once each week in conjunction with the fire alarm test to ensure that the mechanisms are working effectively and the doors close effectively onto their frames.

- The automatic release is fitted as close as possible to the self-closing device in order to reduce the possibility of the door becoming distorted.

Doors used for **means of escape** should be kept unlocked at all times when people are in the building; no door should ever be so fastened that it cannot be easily and immediately opened by people escaping, without the use of a key.

Wherever possible there should be **only one fastening**. However, where more than one fastening cannot be avoided, all but the single emergency fastening should be kept released at all times when people are on the premises.

Where the door:

- may be used at the time of a fire by more than 50 people, or
- is an exit from an area in which fire may develop very rapidly, and
- has to be kept fastened while people are in the building,

it should be fastened only by pressure release devices such as panic latches, panic bolts or pressure pads which ensure that the door can be easily and immediately opened.

Emergency Lighting

Emergency (or safety) lighting should be provided where failure of the normal system would cause problems, e.g. in buildings used **after dark**, or darkened, e.g. cinemas, hospitals, and sections of buildings used for **means of escape**. The purposes of emergency lighting are to:

- Identify the escape route.
- Provide illumination along the route and at exits.
- Ensure that alarm call points and fire-fighting equipment can be easily located.

Emergency lighting

Emergency escape lighting should be provided in those parts of buildings where there is underground or windowless accommodation, core stairways or extensive internal corridors. Generally the need for such lighting will arise more frequently in shops than in factories and offices because of the greater likelihood of people in the building being unfamiliar with the means of escape.

Emergency lighting is only designed to ensure that people **can find their way out of a building**, and so the light requirement is much lower than for normal use. The level of emergency illumination should be related to the level of normal illumination to avoid panic while eyes adapt to the reduced light.

Exit and Directional Signs

Emergency signs must be square or rectangular and must have a **green background with white symbols**. The 'running man' sign (see the following figure) should appear over all fire escape exits, and the 'door and arrow' sign is used to mark escape routes. Text alone, e.g. the word 'Exit', cannot be used, though text can be used to supplement graphics.

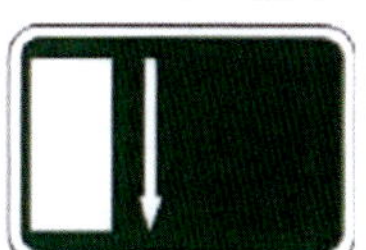

Emergency escape signs

Maintaining Fire Safety in Communal Areas

Access to Competent Advice on Fire Safety Legislation

Landlords and owners of multi-occupancy buildings have a responsibility under the **RRFSO** to maintain fire safety in communal areas. They must appoint a competent person to provide safety assistance, i.e. guidance on the fire safety measures required by the legislation and how they should be implemented. Anyone providing this service to a responsible person must be fully familiar with fire safety requirements in communal areas.

Co-ordination with Other Occupiers

The **RRFSO** also imposes duties on a **responsible person** to "co-ordinate and co-operate" with other responsible persons, either located in the same building or having responsibility for fire safety measures in the building. This may apply where the communal area is part of a development with shops, hotels and other commercial premises, unless there is substantial fire separation between the two and there are no shared escape routes. A key element of this is co-ordination of fire procedures. The difficulties that can arise when a building or its fire safety systems are shared need to be recognised. For example, the detectors needed to operate Automatic Opening Vents (AOVs) in the lobbies to stairways may be part of a fire alarm system covering commercial areas. As the system is common to both, it is important that a single organisation takes responsibility for its testing and maintenance and that there is adequate recourse in contracts and leases to take action if there is a failing on this organisation's part to effect this.

Engaging with Occupants

Landlords and others responsible for managing communal areas should seek to engage with the occupants and communicate a number of vital fire safety messages, including:

- How they can prevent fires in the common parts of the building.
- The importance of maintaining security (making sure doors close behind them when they enter or leave) and being vigilant for deliberate fire setting.
- Not to store petrol, bottled gas, paraffin heaters or other flammable substances inside the building and particularly not in communal areas.
- What action they should take if they discover a fire in the communal area.
- What they must do to safeguard communal escape routes, especially taking care to make sure fire doors self-close properly and are not wedged, tied or otherwise held open.

Occupants in a multi-occupancy building should also be aware of what the fire safety policy is on the use of communal areas and ways they can assist the fire and rescue service by not blocking access when parking, and by keeping fire main inlets and outlets, where provided, clear. Specifically targeted campaigns of leafleting and other initiatives by the landlord to promote fire safety may be necessary to keep the message fresh in people's minds, up to date and relevant to their particular circumstances. It is important that the needs of non-English-speaking, visually impaired and disabled occupants are taken into account.

Key Points

Arrangements for managing fire safety in the communal areas of buildings should include the following:

- Developing a fire policy and appointing someone in the organisation to take overall responsibility for fire safety.
- Making sure someone is designated to provide guidance on fire safety measures required by the **RRFSO**, and supporting this person with help from specialists, where necessary.
- Co-ordinating and co-operating with other occupiers, particularly on issues such as fire procedures.

- Using handbooks, websites and other media to engage with occupants and communicate vital fire safety messages.
- Providing training to ensure safety officers and other occupants have sufficient fire safety awareness.
- Preparing relevant fire procedures and making everyone aware of them.
- Managing the risk from building works, including adopting a 'hot work' permit system.
- Putting in place programmes for routine inspection, testing, servicing and maintenance of fire safety systems and equipment.
- Arranging similar programmes to monitor the condition of other fire safety measures, such as fire-resisting doors.
- Monitoring the common parts through formal inspections and as part of day-to-day activities by staff.
- Carrying out fire risk assessment reviews to monitor standards.
- Putting in place processes for scrutinising planned alterations in order to consider their impact on fire safety.
- Maintaining suitable records.
- Liaising with the fire and rescue service and encouraging occupants to take up the offer of fire-safety checks by the local fire authority.

STUDY QUESTIONS

13. What are the main factors to consider in providing an adequate means of escape?
14. Explain what is meant by a place of relative safety.
15. Identify the four main purposes of smoke-stop fire-resisting doors.
16. Describe when emergency lighting should be provided.

(Suggested Answers are at the end.)

Emergency Evacuation Procedures

IN THIS SECTION...

- Evacuation procedures are part of the fire emergency plan. They cover how the evacuation of the premises should be carried out, where people should assemble after they have left the premises, what the procedures are for checking whether the premises have been evacuated and the identification of key escape routes.
- Evacuation procedures also include the duties of staff who have specific responsibilities if there is a fire and the arrangements for the safe evacuation of people identified as being especially at risk, such as young persons, those with disabilities or lone workers.
- As soon as premises have been evacuated and all personnel are assembled in their pre-arranged safe areas, fire wardens or other appointed persons should carry out a roll call on the basis of a continually updated register.
- Fire wardens (or marshals) are designated people who, in the event of a fire, search and check their allocated area, ensure that all people have left the building, direct those who have not left the building to an appropriate fire exit and safe assembly point and report that their area has been checked and is clear.
- The aim of a Personal Emergency Evacuation Plan (PEEP) is to provide people who cannot get themselves out of a building unaided during an emergency situation, with the necessary information to be able to manage their escape from the building.

Evacuation Procedures and Drills

The specific Government fire guidance for offices and shops referenced earlier in this Learning Outcome suggests that a fire emergency plan should cover:

Identification of key escape routes

- Method by which people will be warned if there is a fire.
- What staff should do if they discover a fire.
- How the evacuation of the premises should be carried out.
- Where people should assemble after they have left the premises and procedures for checking whether the premises have been evacuated.
- Identification of key escape routes, how people can gain access to them and escape from them to a place of total safety.
- Arrangements for fighting fire.
- The duties and identity of staff who have specific responsibilities if there is a fire:
  - The incident controller and deputy incident controller.
  - Fire marshals and deputy fire marshals.
  - Roll callers.
  - Supervisors.
- Arrangements for the safe evacuation of people identified as being especially at risk, such as young persons, those with disabilities or lone workers.
- Any machines/processes/appliances/power supplies that need to be stopped or isolated if there is a fire.

- Specific arrangements for high fire-risk areas.
- Contingency plans for when fire safety systems, such as evacuation lifts, fire detection and warning systems, sprinklers or smoke control systems, are out of order.
- How the fire and rescue service and any other necessary services will be called and who will be responsible for doing it.
- Procedures for meeting the fire and rescue service on their arrival and notifying them of any special risks, e.g. the location of highly flammable materials.

To help people understand the emergency plan (and also the fire risk assessment, as discussed earlier in this Learning Outcome), it is also a good idea to include a building plan with fire-related features marked on it, such as:

- Positions of escape doors.
- Positions of escape routes.
- Fire-resisting constructions.
- Location of refuges and lifts designed to assist those who may need assistance.
- Methods of fighting fire, e.g. numbers and type of fire extinguishers at each location.
- Location of manually operated fire-alarm call points.
- Location of control equipment for fire alarms.
- Location of emergency lighting and exit route signage.
- Location of high fire-risk areas and processes.
- Location of automatic fire-fighting systems such as sprinklers.
- Location of main electrical supply/isolation switch, main water shut-off valve, main gas and oil shut-off valves.

Training

Training in fire safety procedures is necessary for everybody. It is essential that key personnel who are responsible for implementing safety procedures are given adequate training in order to perform their duties. The skill is in identifying the level of training required. This can range from induction training for new employees, to more specialised training for members of staff with specific safety responsibilities, for example, fire wardens and people dealing with specialist processes or equipment. Training must be monitored and updated as necessary (e.g. when circumstances change in premises or staffing) and should be modified to suit individual requirements.

Keeping records of the training given to individuals will not only ensure that all employees are made aware of their responsibilities, but also provides proof of provision of adequate training. The records will also be invaluable in dealing with investigations of any incidents or accidents. Carrying out and recording the results of regular risk assessments can also help identify training needs.

Training should be specific to the particular premises and all staff should receive training at sufficiently regular intervals to ensure that existing members of staff are reminded of the action to take, and that new staff are made aware of the fire routine for the premises. Training should be given at least once every 12 months, but in some circumstances where there is high turnover of staff, a high fire risk, or material changes which significantly affect the fire safety arrangements, training may have to be more frequent. Instruction and training should be based on written procedures and should be appropriate to the duties and responsibilities of the staff.

It is particularly important that all staff (including those casually employed) should be shown the means of escape and told about the fire routine as soon as possible after they start work. It is also necessary to ensure that occasional workers, those on shift duties and others who work in the premises are similarly instructed. Special consideration should be given to any employees with language difficulties or with any disabilities which may impede their understanding of the information.

All **staff** including those appointed as **roll callers** should receive fire safety training at induction and at yearly intervals thereafter. Training should cover:

- What to do on discovering a fire.
- How to raise the alarm and what happens then.
- What to do on hearing the fire alarm.
- The procedures for alerting contractors and visitors including, where appropriate, directing them to exits.
- The arrangements for calling the fire and rescue service.
- The evacuation procedures for everyone on the premises to reach an assembly point at a place of total safety.
- The location and, when appropriate, the use of fire-fighting equipment.
- The location of escape routes, especially those not in regular use.
- How to open all emergency exit doors.
- The importance of keeping fire doors closed to prevent the spread of fire, heat and smoke.
- Where appropriate, how to stop machines and processes and isolate power supplies in the event of a fire.
- The reason for not using lifts (except those specifically installed or nominated, following a suitable fire risk assessment).
- The safe use of, and risks from storing or working with, highly flammable and explosive substances.
- The importance of general fire safety, which includes good housekeeping.
- The items listed in the **emergency plan**.
- The importance of fire doors and other basic fire-prevention measures.
- The importance of reporting to the assembly area.
- Exit routes and the operation of exit devices, including physically walking these routes.
- General matters such as smoking policy and permitted smoking areas or restrictions on cooking other than in designated areas.
- Assisting disabled persons where necessary.

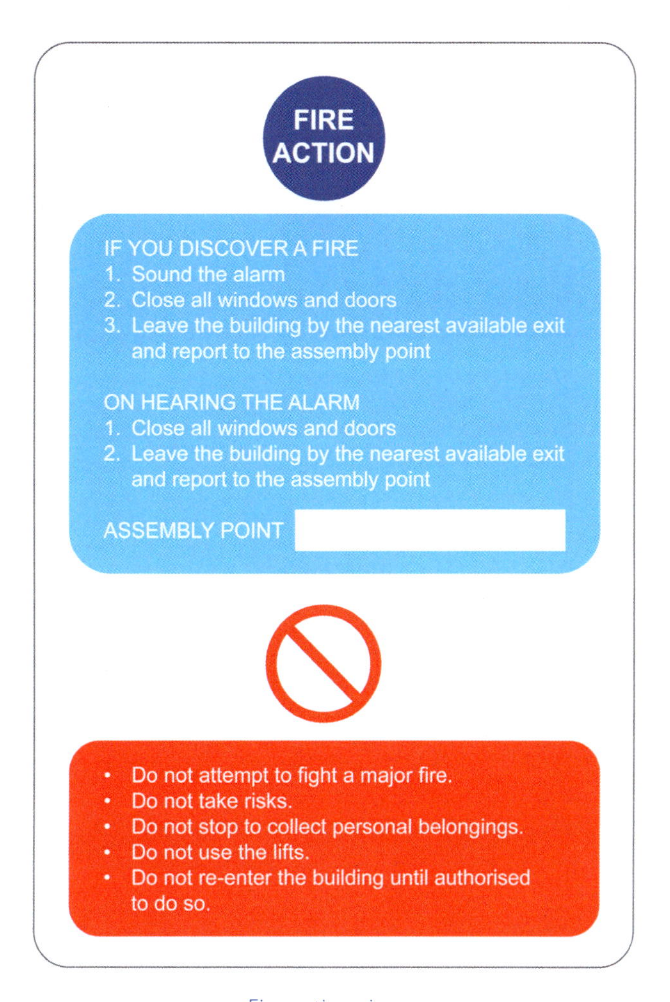

Fire action sign

Fire marshals/wardens, **deputy fire marshals/wardens**, **incident controllers** and **deputy incident controllers** should receive additional training covering:

- Detailed knowledge of the fire safety strategy of the premises.
- Awareness of human behaviour in fires.
- How to encourage others to use the most appropriate escape route.
- How to search safely and recognise areas that are unsafe to enter.
- The difficulties that some people, particularly if disabled, may have in escaping and any special evacuation arrangements that have been pre-planned.
- Additional training in the use of fire-fighting equipment.
- An understanding of the purpose of any fixed fire-fighting equipment such as sprinklers or gas flooding systems.
- Reporting of faults, incidents and near misses.

A fire drill should be carried out **at least once and preferably twice a year**, simulating conditions in which one or more of the escape routes from the building is obstructed. During the drills the fire alarm should be operated by a member of staff who is told of the supposed outbreak, and thereafter the fire routine should be rehearsed as fully as circumstances allow.

The training and instruction given should be recorded in a log or other suitable record, which should be available for inspection. The following are examples of matters which should be included in a fire record-keeping book:

- Date of the instruction or exercise.
- Its duration.
- Name of the person giving the instruction.
- Names of the persons receiving the instruction.

On all premises, one person should be responsible for organising fire instruction and training, and in larger premises, a person or persons should be nominated to co-ordinate the actions of the occupants in the event of a fire.

Printed notices should be displayed at conspicuous positions in the building, stating in concise terms the action to be taken on discovering a fire or on hearing the fire alarm. The notices should be permanently fixed in position and suitably protected to prevent loss or defacement. Written instructions may be supplemented by advice in pictogram form.

Alarm Evacuation and Roll Call

With any building in which a great number of people are present, the procedure must not allow panic as it will slow evacuation and may cost lives. The evacuation must be orderly and handled in such a manner that the escape facilities, e.g. capacity of escape routes, can handle the number of people using them.

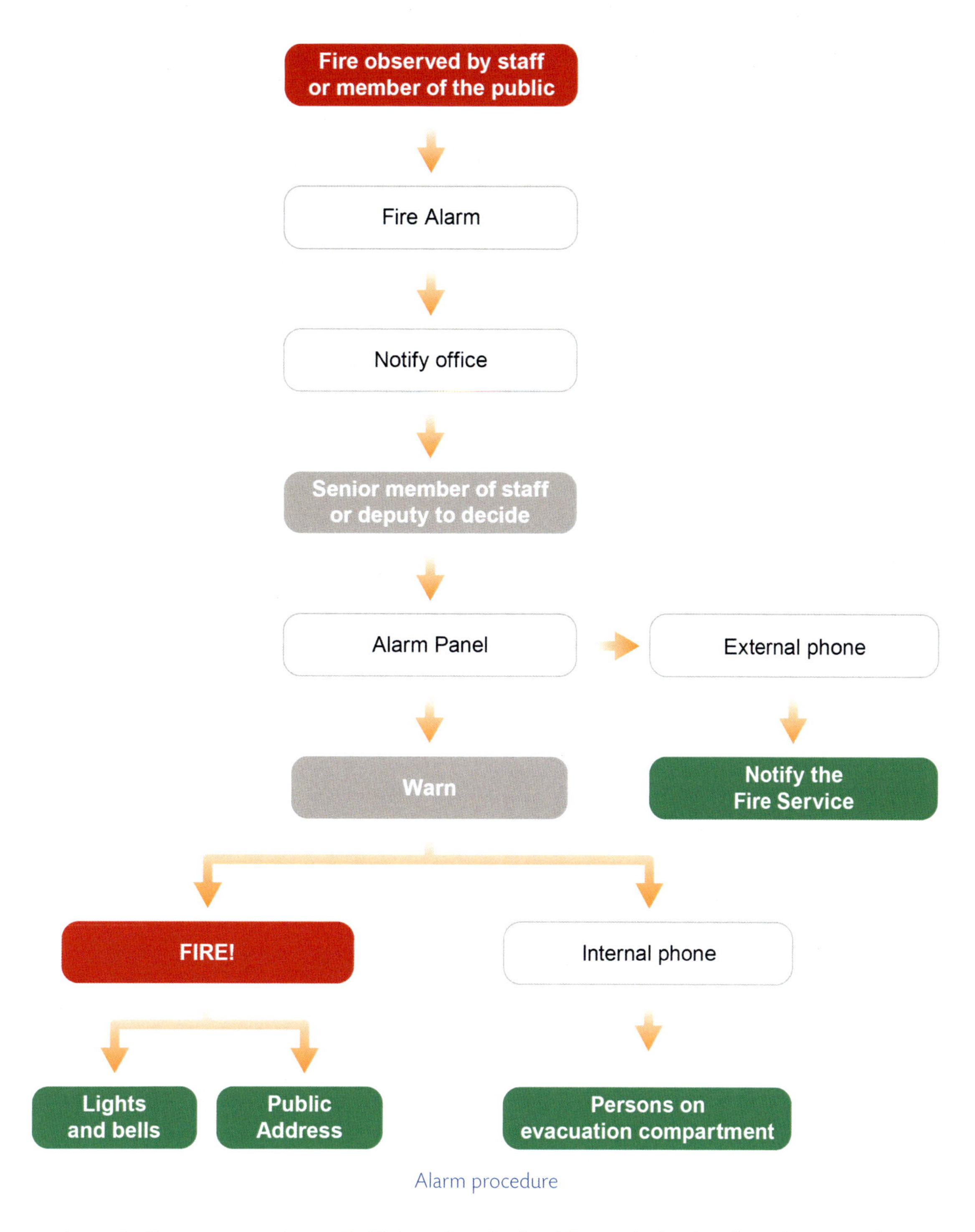

Alarm procedure

A massed crowd will cause congestion and is likely to slow the flow. The method to be adopted is a **zoned evacuation**, i.e. a compartment at a time. The compartments must be evacuated in sequence - the one at highest risk first. Sounding an alarm throughout the building would not achieve this and may slow total evacuation.

Staff duties (shops, public buildings):

- Detailed members of staff shepherd the public to the escape route.
- Senior members of staff search all the floors, WCs, etc.
- Members of staff keep exits open and clear.

- A responsible member should meet the fire service on arrival and inform them of all relevant details.
- On leaving the premises ensure all doors and windows are closed.

As soon as the premises have been evacuated and all personnel are assembled in their pre-arranged safe areas, fire wardens or other appointed persons must carry out a **roll call** on the basis of a continually updated register (the compilation of which will be a line management function). The roll must include staff, visitors, contractors, etc.

If anyone is not accounted for, the fire officer in charge must be notified as soon as the fire services arrives.

In the case of premises where there is random public access (shops, etc.), it may be necessary for the fire services to undertake a search.

Provision of Fire Wardens and Their Role

Fire marshals, also known as 'fire wardens':

- Are appointed to assist with the evacuation procedure.
- Require more comprehensive training (as above).
- Roles include:
  - **Helping** any members of the public, visitors and/or disabled persons leave the premises.
  - **Checking** designated areas to ensure everyone has left.
  - **Using** fire-fighting equipment.
  - **Liaising** with the fire and rescue service.
  - **Shutting down** vital or dangerous equipment.
  - Performing a supervisory/**managing** role in any fire situation.

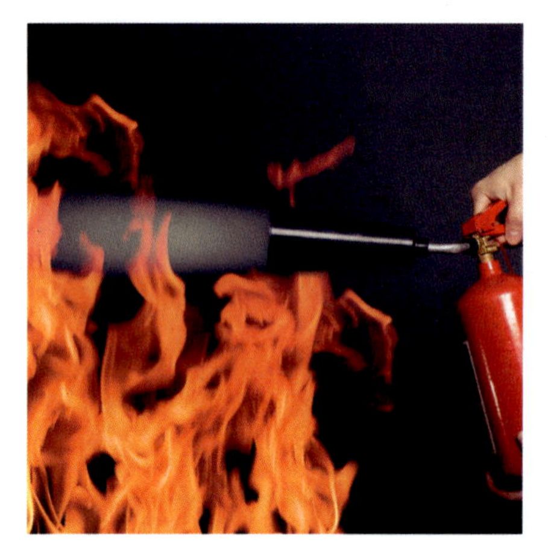

A fire warden's role includes using fire-fighting equipment

TOPIC FOCUS

Fire Marshal Systems

The operation of the system involves:

- The building being split into small areas of responsibility.
- Each area is allocated to a specific fire marshal.
- The marshals are designated people who, in the event of a fire:
  - Search and check their allocated area.
  - Ensure that all people have left the building.
  - Direct those who have not left the building to an appropriate fire exit and safe assembly point.
  - Report that their area has been checked and is clear.

The benefits of the system include:

- The use of trained persons who are familiar with the premises to evacuate other people who may not be familiar with the premises.
- Marshals can compensate for any adverse human behaviour which might hinder or delay the evacuation.

Personal Emergency Evacuation Plans (PEEPs)

The aim of a Personal Emergency Evacuation Plan (PEEP) is to provide people who cannot get themselves out of a building unaided during an emergency situation, with the necessary information to be able to manage their escape from the building.

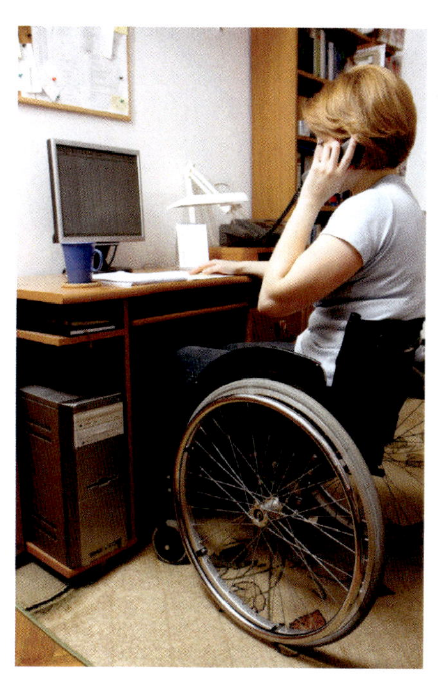

Disabled persons may require a Personal Emergency Evacuation Plan (PEEP)

Individuals with a **hearing** impairment may take longer to be alerted, and the **visually** impaired may have greater difficulty finding their way out of a building. These issues need to be considered in the emergency evacuation plan for those persons.

To execute the evacuation plan, special aids may be needed, such as:

- **Personal trembler alarms** which 'vibrate' at the same time as the alarm.
- A **buddy system** where someone is allocated the task of helping the person with a sensory impairment.
- **Visual alarms** for the **hearing** impaired such as a flashing beacon.
- **Tactile/Braille** signs for the **visually** impaired, providing the person can locate the sign and is able to read Braille.

Persons with severe **mobility** impairment may require assistance during a fire evacuation from building design elements and accessories such as:

- **Evacuation lifts** designated for the use of the disabled which continue to operate during the fire.
- **Evacuation chairs**, lightweight, manoeuvrable, specialised wheeled chairs for evacuation down stairs and along corridors.
- **Refuges**, fire-protected areas (minimum 30 minutes' fire resistance) which offer **temporary** relative safety where a disabled person has to wait for assistance for full evacuation or needs to rest. They are generally located within a protected stairwell and close to an evacuation lift/chair.

STUDY QUESTIONS

17. Identify the issues to be covered during a training session on emergency procedures.
18. Explain the operation and benefits of a fire marshal system.
19. Describe the special aids that might be required in order to execute the PEEP for persons with severe sensory impairment.

(Suggested Answers are at the end.)

Summary

Legal Requirements

We have:

- Examined the regulatory powers of a fire authority with respect to fire safety, particularly the powers of inspectors to serve Alterations Notices, Enforcement Notices and Prohibition Notices.
- Considered dual enforcement, which is the sharing of powers between fire and rescue authorities and enforcing authorities for health and safety at work.
- Outlined the requirements of the **Regulatory Reform (Fire Safety) Order 2005 (RRFSO)**, including the duties of the responsible person.
- Outlined the purpose of the **Building Regulations 2010** Approved Document B.

Identification of Hazards and the Assessment of Risk from Fire

We have:

- Considered fire hazards and the assessment of risk.
- Outlined the five steps to fire risk assessment:
  - Identify fire hazards.
  - Identify people at risk.
  - Evaluate, remove, reduce and protect from risk.
  - Record, plan, inform, instruct and train.
  - Review.

Fire Detection and Alarm Systems

We have:

- Examined the common fire detection and alarm systems and procedures, including types of fire detector - ionisation, optical, radiation and heat detectors - which are activated by changes in the physical environment.
- Noted that alarm systems fall into three main categories: manual, manual/electric and automatic.

Fixed and Portable Fire-Fighting Equipment

We have:

- Discussed the factors in the design and application of fixed fire fighting systems and equipment including sprinklers, drenchers, total flood systems and deluge systems.
- Examined the classification of fires and identified the main types of extinguishing media used in fixed installations such as water, foam, carbon dioxide, dry powder, wet chemical and other vaporising liquids.
- Considered that portable fire-fighting equipment should be sited in an easily seen and reached position and regularly inspected and maintained. We also noted there should be suitable training provided for persons required to use it.
- Highlighted the potential environmental damage that can result from contaminated fire-fighting water run-off,

meaning that sites should consider the polluting effects of fire in their emergency plans.

Means of Escape

We have:

- Identified the factors which require particular consideration for the provision of safe means of escape, including the nature/vulnerability of the occupants, the number of people attempting to escape, the distance they may have to travel to reach a place of safety, and the size and extent of the 'place of safety'.
- Considered the general requirements for a safe means of escape, including travel distances, stairs, passageways and doors, emergency lighting and signs.
- Outlined the requirements for maintaining fire safety in communal areas, including access to competent advice on fire safety legislation, co-ordination with other occupiers and engaging with occupants to communicate fire safety messages.

Emergency Evacuation Procedures

We have:

- Recognised that evacuation procedures are a part of the emergency plan which covers:
  - How evacuation should be carried out, where people should assemble, how evacuation of the premises is checked and identifies the key escape routes. These factors should be tested using emergency drills.
  - The duties of staff who have specific responsibilities if there is a fire, including arrangements for the safe evacuation of people especially at risk and taking roll calls.
- Identified fire wardens (or marshals) as designated people who, in the event of a fire, search and check their allocated area, ensure that all people have left the building, direct those who have not left the building to an appropriate fire exit and safe assembly point and report that their area has been checked and is clear.
- Considered the requirement to provide PEEPs for anyone who cannot get themselves out of a building unaided during an evacuation.

Learning Outcome 11.5

NEBOSH National Diploma for Occupational Health and Safety Management Professionals

ASSESSMENT CRITERIA

- Describe the risks and controls inherent in industrial chemical processes and hazardous environments, including the storage, handling and transport of dangerous substances and planning for emergencies.

LEARNING OBJECTIVES

Once you've studied this Learning Outcome, you should be able to:

- Outline the main physical and chemical characteristics of industrial chemical processes.
- Outline the main principles of the safe storage, handling and transport of dangerous substances.
- Outline the main principles of the design and use of electrical systems and equipment in adverse or hazardous environments.
- Explain the need for emergency planning, the typical organisational arrangements needed for emergencies and relevant regulatory requirements.

Contents

Industrial Chemical Processes

IN THIS SECTION...

- The rate of a chemical reaction will usually increase with **temperature** and **pressure**.
- A **catalyst** will affect the rate of a chemical reaction without being changed itself.
- **Exothermic** reactions are accompanied by the **evolution of heat**, e.g. combustion of propane.
- A **runaway reaction** is an exothermic reaction where the heat generated continues to increase the temperature, accelerating the reaction out of control, e.g. Bhopal in 1984.

Introduction to Industrial Chemical Processes

Industrial chemical processes, such as those conducted at an oil refinery or at a polyethylene manufacturing plant, often involve the storage and handling of large quantities of chemicals.

Oil refineries involve storage and handling of large quantities of chemicals

These chemicals are sometimes distilled (i.e. boiled and condensed to separate mixtures) or 'cracked' in the refinery process and are frequently reacted together to form products. This happens inside reactor vessels, often at high temperatures and pressures.

The reactive chemical hazards associated with the process are very significant and some of the worst global industrial disasters have involved reactive chemistry and the catastrophic loss of control of chemical reactions.

These disasters have often involved multiple fatalities, destruction of plant and the release of significant quantities of toxic material into the local environment and the ongoing human and environmental health effects have been long lasting and profound. Ill health in the adult population still occurs in Bhopal, India, where one of the largest and most notorious chemical releases occurred on the night of 2-3 December 1984 at the Union Carbide pesticide plant.

Effects of Temperature, Pressure and Catalysts

Temperature

The rate at which a chemical reaction occurs is **temperature sensitive**. The higher the temperature, the faster the rate of reaction. This occurs because the molecules making up a chemical mixture (whether in liquid or vapour/gas form) move much more quickly at high temperatures, colliding and reacting more frequently. In most cases, an increase of 10°C roughly doubles the reaction rate. So, if a chemical reaction proceeds at rate R at 0°C it will run at $2R$ at 10°C; $4R$ at 20°C; $8R$ at 30°C; $16R$ at 40°C; etc.

It is not difficult to see that large increases in temperature will therefore have correspondingly significant effects on the rate of reaction.

Pressure

The rate at which a chemical reaction takes place is also **pressure sensitive** and applies equally to both gases and vapours. When gases and vapours are compressed, the molecules are closer together and will frequently collide. This molecule collision forms a reaction and, depending on the pressure applied, we know that a mixture of gases/vapours at low pressure will react at a slower rate than the same mixture that has been compressed at a higher pressure.

This effect is compounded by the fact that the act of compression generates heat. A compressed mass of gas will be hotter than the same volume of gas that has been expanded. This is why compressing air in a bicycle pump generates heat and why CO_2 extinguishers get very cold when being discharged.

Catalysts

DEFINITION

CATALYST

Any agent (usually a substance) which, when added in very small quantities, notably affects the rate of a chemical reaction without itself being consumed or undergoing a chemical change. Most catalysts accelerate reactions but a few retard them ('negative catalysts' or 'inhibitors').

Catalysts are added to chemical reactions to speed up (or slow down) the reaction rate. They can be highly specific in their application, and are essential in virtually all industrial chemical reactions, especially petroleum refining and synthetic organic chemical manufacturing.

The activity of a solid catalyst is often centred on a small fraction of its surface, so the number of active points can be increased by breaking the catalyst down into a fine powder form or adding promoters which increase the surface area in some way, e.g. by increasing porosity. Catalytic activity is decreased by substances that act as poisons which clog and weaken the catalyst surface. The catalytic converter used as a part of the exhaust of a modern car petrol engine would be poisoned by the lead in leaded petrol if that were used as a fuel.

Heat of Reaction

Exothermic Reactions

When a chemical reaction takes place it will either consume or produce heat. If the chemical reaction **consumes** heat from the local environment, it is known as an **endothermic** reaction. For example, if you dissolve ammonium nitrate (the fertiliser we encountered earlier in the unit) in water, the water will get cold - the process is endothermic.

If a chemical reaction **produces** heat then it is known as **exothermic**. The combustion reaction is exothermic in nature; meaning that if you burn, for example, methane gas in air, you will produce carbon dioxide, water and **heat**.

Exothermic reactions present very significant risks for obvious reasons. If the heat produced by the reaction is significant, it may be possible that the reactant chemicals or their products may be heated to their auto-ignition temperature and a fire or explosion will result.

Materials in close proximity to the reacting chemicals may also be heated to auto-ignition or, the chemical reaction may cause spontaneous combustion where no external ignition/heat source is required. Some very significant industrial fires have started as a result of uncontrolled exothermic reactions taking place. For example, the fire that started at Allied Colloids Ltd chemical storage facility near Bradford in 1992, was caused by incompatible chemicals being stored adjacent to one another and mixing together due to spillage.

MORE...

Search for Allied Colloids Ltd fire at Low Moor, Bradford. The HSE website has information about this incident. The Institute of Chemical Engineers' (iChemE) website has access to the official inquiry report.

Even very small quantities of reactive chemicals can produce heat and combustion when mixed together in the right proportions.

We know that the rate at which chemical reactions take place depends on the temperature of the reactants and, in fact, the rate of reaction increases as the temperature of the chemicals rises. A 10°C increase in temperature typically causes a doubling of reaction rate, but an exothermic reaction also **produces** heat as it occurs. This means that some chemical reactions will produce heat that will then make the reaction speed up; a positive feedback loop is then created that leads to a thermal runaway reaction.

Runaway Reactions

A runaway reaction occurs when chemicals are mixed together and an exothermic reaction takes place in an uncontrolled manner:

- the reacting chemicals produce heat (remember the reaction is exothermic);
- this causes the reaction rate to increase (remember that a 10°C temperature increase normally doubles the rate of reaction);
- this causes the production of more heat at a faster rate;
- that causes a faster reaction rate;
- that increases the rate of heat production;
- that causes a faster reaction rate; etc.

The end result will be a total loss of control of the rate of reaction, excessive heat production and excessive pressure build up as a result of that heating. This will be followed by loss of containment of the reacting chemicals and potentially fire and explosion if the chemicals are flammable/combustible and auto-ignition temperatures are achieved.

This usually occurs in a reactor vessel or storage tank. The vessel may be a pressure vessel. The reactor vessel may already be heated to a high temperature in order for the ordinary chemical reactions to take place at a proper rate. There may or may not be a catalyst present.

MORE...

The US Chemical Safety and Hazard Investigation Board (CSB) have made an excellent video about the fire at T2 Laboratories in Florida 2007 that is available online.

See:

www.youtube.com/watch?v=C561PCq5E1g or search for "T2 laboratories".

Causes of Incidents

TOPIC FOCUS

The conditions that might give rise to a '**runaway reaction**' include:

- A strongly exothermic reaction process.
- Inadequate provision of cooling.
- Catalysis by contaminants.
- Lack of temperature detection and control.
- Excessive quantities of reactants in the reaction vessel.
- Failure of mixing or agitation.

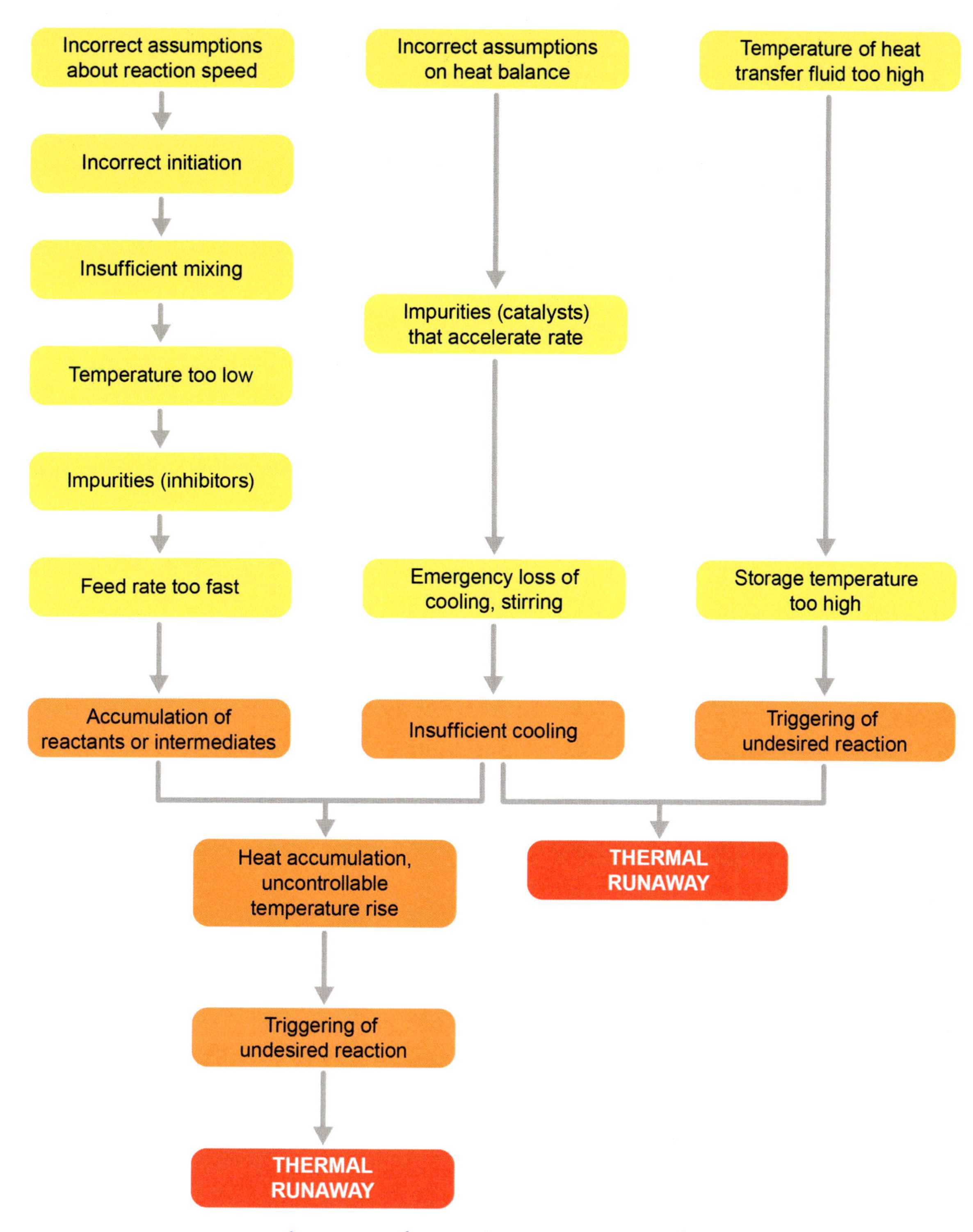

Some causes of runaway in reactors or storage tanks

The Scale of the Reactant Mass

The scale with which a reaction is carried out can have a significant effect on the likelihood of runaway. The heat produced increases with the volume of the reaction mixture, whereas the heat removed depends on the surface area available for heat transfer. As scale, and the ratio of volume to surface area, increases, cooling may become inadequate. This has important implications for scale-up of processes from the laboratory to production and should be taken into account, particularly when modifying a process to increase the reaction quantities.

Effects of Thermal Runaway

A runaway exothermic reaction can have a range of results from the boiling over of the reaction mass, to large increases in temperature and pressure that can lead to an explosion. Such violence can cause blast and missile damage. If flammable materials are released, fire or a secondary explosion may result. Hot liquors and toxic materials may contaminate the workplace or generate a toxic cloud that may spread off-site.

There can be serious risk of injuries, even death, to plant operators, and the general public and the local environment may be harmed. At best, a runaway causes loss and disruption of production; at worst, it has the potential for a major accident.

Examples of Exothermic and Runaway Reactions

Exothermic Reactions

As we saw earlier in the unit, the combustion reaction is an exothermic chemical reaction. For example, propane gas reacts exothermically with oxygen. It will burn in excess oxygen with the generation of heat to form water and carbon dioxide.

$$\text{propane + oxygen} \rightarrow \text{carbon dioxide + water + heat}$$

$$C_3H_8 + 5O_2 \rightarrow 3CO_2 + 4H_2O$$

To proceed, this reaction will require heat to start (an ignition source) unless the propane/air mixture is at or above its auto-ignition temperature (460°C).

Many exothermic chemical reactions can occur at low temperatures and then raise the temperature of the reactants or surrounding materials to auto-ignition temperatures where combustion will occur. Examples include:

- Mixtures of potassium permanganate (an oxidising substance) and glycerin.
- Sawdust mixed with oil-based paint dust (as might occur during paint stripping/renovation work).
- Haystacks - especially if the hay is damp and the stack is large.
- Compost heaps (again, especially if it is damp and large).
- Cotton rags soaked in oil (especially if put in a metal bin in large quantities).

Some exothermic reactions may be so violent they become explosions. Ammonium nitrate mixed with fuel oil can explode if mixed together in the correct proportions and detonated. This mixture of fertiliser and oil is used as a common explosive (ANFO) in mining and quarrying.

Runaway Reaction

Bhopal, 1984

Methyl Isocyanate (MIC) is an extremely toxic and unstable substance and even in very small quantities is fatal; it is also volatile and unstable at higher temperatures and will break down exothermically to give out large amounts of heat if contaminated. For safe handling, it must be maintained at about 0°C.

On the night of 2-3 December 1984 at the Union Carbide India Limited (UCIL) pesticide plant in Bhopal, Madhya Pradesh, India, during a routine cleaning operation, maintenance workers inadvertently allowed water to enter an MIC storage tank (Tank E610), causing an exothermic runaway reaction to occur. The MIC boiled and toxic vapour was expelled through the bursting disc vent. A scrubber and flare stack, which might have been able to deal with the release, were shut down at the time so the MIC vapour was discharged directly to atmosphere. The wind carried the plume over the adjacent city of Bhopal where more than half a million people were exposed to MIC gas and other chemicals. The local government put the immediate death toll at over 3,500, but more recent estimates have driven that figure up to perhaps 8,000. Many victims suffered long-term health effects from the event and many continue to do so.

Bhopal is widely recognised as the world's worst industrial accident.

Methods of Controlling Exothermic and Runaway Reactions

An analysis of thermal runaways has indicated that incidents occur because of inadequacies in:

- understanding of process and thermal chemistry;
- design for heat removal;
- control systems and safety systems; and
- operational procedures, including training.

These are some of the key factors that should be considered when defining a safe process.

Control of process safety when reactive hazards are present depends on detailed knowledge of the reaction chemistry. Operations of this kind should only be entrusted to highly skilled and competent industrial chemists knowledgeable in the dangers involved and the precautions to be taken.

Important factors in preventing such thermal-runaway reactions are mainly related to the **control of reaction velocity and temperature** within suitable limits. These will include considerations such as:

- Ensuring a comprehensive hazard and operability study is conducted before the process is built or put into production.
- Adequate cooling capacity in both liquid and vapour phases of a reaction system. Cooling mechanisms must be reliable; perhaps with back-up cooling available.
- The proportions of the reactants and the rates of addition must be carefully controlled (allowing for an induction period whilst the reaction gets underway).
- Adequate agitation and mixing in the reaction vessel to ensure that local hot spots or areas of inadequate mixing do not occur. Design of agitators and the reliability of mechanisms must be considered.
- The use of an inert atmosphere to prevent fire.
- The provision of adequate pressure relief (e.g. bursting discs) so that in the event of a runaway reaction the reactor vessel does not catastrophically fail but excess pressure is safely vented to atmosphere.
- The use of solvents as diluents and to reduce viscosity of the reaction medium.
- The control of reaction or distillation pressure.

Many chemical processes require equipment designed to rigid specifications, with sophisticated automatic control and safety devices. With some reactions, it is important to provide protection against failure of cooling systems, agitation, control, or safety instrumentation, etc. This requires that the highest-specification equipment and systems are selected and fitted; that back-up equipment and systems are fitted (for redundancy) and that they are maintained in good working order.

STUDY QUESTIONS

1. In general terms, how is the rate of a chemical reaction affected by temperature?
2. What are the two key factors in controlling a thermal runaway?

(Suggested Answers are at the end.)

Storage, Handling and Transport of Dangerous Substances

IN THIS SECTION...

- Employers and the self-employed have a duty under **DSEAR** Regulation 5 to assess, eliminate or reduce risks from dangerous substances.
- The amount of flammable solids or liquids held in a workplace should be limited. Large quantities of flammable substances should be held in suitable stores. Storage may be in:
  - Bulk storage tanks.
  - Intermediate storage vessels.
  - Drums.
- Incompatible materials should be identified and segregated to ensure that they are not allowed to mix and react if a spillage or release occurs.
- Means of controlling spillage should be provided where materials are stored, e.g. an impervious sill or low bund. During filling and transfer, release of vapours should be prevented or kept to an absolute minimum.
- The risks arising from the storage and handling of dangerous substances may be reduced by:
  - Measures to prevent loss of containment or static charge generation during flow through pipelines.
  - Principles of filling and emptying containers.
  - Principles of dispensing, spraying and disposal of flammable liquids.
  - The dangers of electricity in hazardous areas and the identification of 'zones' which relate to the presence, or possible presence, of flammable atmospheres and suitable electrical equipment that may be used in those zones.
- Carrying goods by road or rail involves the risk of traffic accidents. If the goods carried are dangerous, there is also the risk of an incident, such as spillage of the goods, leading to hazards such as fire, explosion, chemical burn or environmental damage. These risks may be minimised by:
  - Adherence to the key safety principles for loading and unloading of tankers and tank containers.
  - Correct labelling of vehicles and packaging of substances.
  - The importance of driver training and the role of the Dangerous Goods Safety Adviser under the **Carriage of Dangerous Goods and Use of Transportable Pressure Equipment Regulations 2009**.

Dangerous Substances Risk Assessment

Regulation 5 of **DSEAR** places duties on employers and the self-employed, to assess and eliminate or reduce risks from dangerous substances.

The Approved Code of Practice that accompanies the Regulations provides practical and pragmatic advice on complying with **DSEAR** and involves:

Assessing Risks

Before work is carried out, risks presented by dangerous substances must be assessed. The assessment should be carried out by a competent person and must take into account:

- The dangerous substances in the workplace (the hazards).
- Who might be harmed, together with the likelihood and severity of the consequences.
- Those inexperienced or vulnerable employees who may be at increased risk and also contractors, visitors and even those who are not lawfully on the premises.
- The work activities involving those substances must be assessed and the ways in which those substances and work activities could harm people.

This exercise should help employers decide what they need to do to eliminate or reduce the risks from the dangerous substances that exist in their workplace.

As in other risk assessments - if there is no risk to safety, or the risk is trivial, then no further action is required. But, if there are risks, then employers must look at what else needs to be done to reduce or eliminate those risks.

Note: It has to be remembered that Regulation 5 also covers the duty to risk assess non-routine maintenance and related higher risk activities such as process scale up, maintenance, repair and modifications - where, in many instances the lack of risk assessment has led to catastrophic plant failures, fires and explosions (see Bhopal).

Preventing or Controlling Risks

The employer must then put control measures in place to eliminate risks from dangerous substances, or reduce them so far as is reasonably practicable.

Where it is not possible to eliminate the risk completely, employers must take measures to control risks and reduce the severity (mitigate) the effects of any harmful event.

According to the HSE, the best solution is to eliminate the risk completely by replacing the dangerous substance with another substance, or using a different work process. This is called 'substitution' in the Regulations.

In practice, this may be difficult to achieve – but it may be possible to reduce the risk by using a less dangerous substance, e.g. replacing a low flashpoint liquid with a higher flashpoint one.

There will be times when it may not be possible to replace the dangerous substance at all and the assessor, together with the employer, will have to think about the balance of risk against the time, trouble, cost and effort of eliminating or reducing the risk (so far as is reasonably practicable).

Control Measures

Employers must look at the current control measures that are already in place and, where the risk cannot be eliminated, the regulations require control measures to be applied in the following priority order:

- Reduce the quantity of dangerous substances to a minimum.
- Avoid or minimise releases of dangerous substances.
- Control releases of dangerous substances at source.
- Prevent the formation of a dangerous atmosphere.
- Collect, contain and remove any releases to a safe place (e.g. through ventilation).
- Avoid ignition sources.

- Avoid adverse conditions (e.g. exceeding the limits of temperature or control settings) that could lead to danger).
- Keep incompatible substances apart.

These control measures should be consistent with the risk assessment and appropriate to the nature of the activity or operation.

Mitigation

In addition to control measures, **DSEAR** requires employers to put mitigation measures in place. These measures should be consistent with the risk assessment and appropriate to the nature of the activity or operation and include:

- Reducing the number of employees exposed to the risk.
- Providing plant that is explosion-resistant.
- Providing plant that is corrosion-resistant.
- Providing explosion suppression or explosion relief equipment.
- Taking measures to control or minimise the spread of fires or explosions.
- Providing suitable personal protective equipment.

Preparing Emergency Plans and Procedures

Arrangements must be made to deal with emergencies. These plans and procedures should cover safety drills and suitable communication and warning systems and should be in proportion to the risks. If an emergency occurs, workers tasked with carrying out repairs or other necessary work must be provided with the appropriate equipment to allow them to carry out this work safely.

The information in the emergency plans and procedures must be made available to the emergency services to allow them to develop their own plans if necessary.

Providing Information, Instruction and Training for Employees

Employees must be provided with relevant information, instructions and training. This includes:

- The dangerous substances present in the workplace and the risks they present including access to any relevant safety data sheets and information on any other legislation that applies to the dangerous substance.
- The findings of the risk assessment and the control measures put in place as a result (including their purpose and how to follow and use them).
- Emergency procedures.

Information, instruction and training need only be provided to other people (non-employees) where it is required to ensure their safety. It should be in proportion to the level and type of risk.

The contents of pipes, containers, etc. must be identifiable to alert employees and others to the presence of dangerous substances. If the contents have already been identified in order to meet the requirements of other law, this does not need to be done again under **DSEAR**.

Storage Methods and Quantities

In this section, the phrase '**dangerous substances**' will be used to mean substances that have been mentioned in previous Learning Outcomes:

- Explosives.
- Oxidising substances.
- Flammable liquids, gases and solids.
- Explosible dusts.

Intermediate bulk containers

These substances are often stored on industrial sites in intermediate or bulk containers of various types. In the first part of this section we will look at some of the safety issues associated with bulk and intermediate storage.

The principal risks associated with the storage of dangerous substances are **fire and explosions**; however, consideration must also be given to other risks, such as:

- **Theft** - some of these substances, such as fuels, have a high monetary value; others might be stolen because they can be used for terrorism or other criminal purposes.
- **Arson** - some of these substances and their containers are relatively easy to set alight.
- **Spillage, leakage and loss of containment** - many of these substances are capable of causing significant damage to the environment if released to atmosphere, into local watercourses or into the ground.

Bulk Storage

Bulk storage usually refers to the storage of quantities of dangerous substances in quantities greater than 1,000 litres or 1 metric tonne. The principal hazards associated with the bulk storage of highly flammable liquids are fire and explosion arising either from:

- Ignition of flammable vapours arising from the stored liquid.
- Vapour escaping from the installation due to equipment failure during process operations.
- Exposure of the storage installation to heat from fire in the vicinity.

The storage of highly flammable liquids in fixed bulk tanks is preferable to storage in drums, but this may not always be practicable.

Storage above ground and in the open air is better, because leaks will usually be visible and therefore detected fairly quickly. Moreover, vapours arising from minor leaks will normally be dissipated by natural ventilation. Cleaning, repair work and modifications can also be more easily carried out. Bulk storage tanks should not be located inside or on the roof of a building. Refer to the following figures for more information.

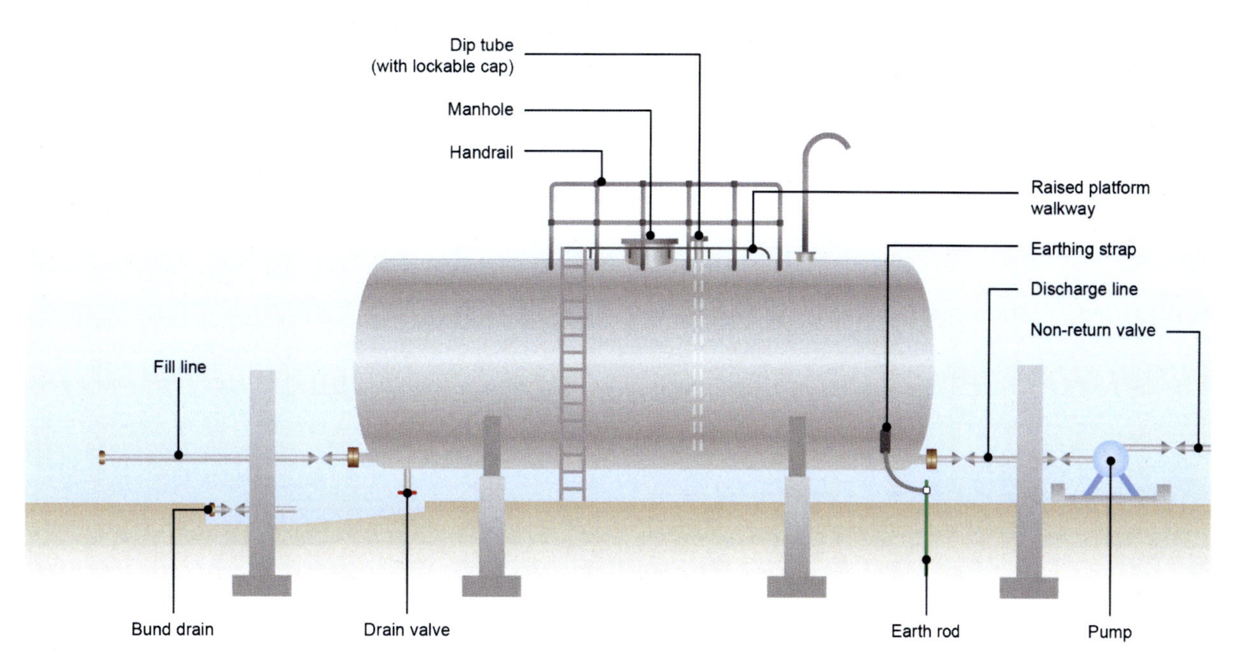

Typical horizontal storage tank

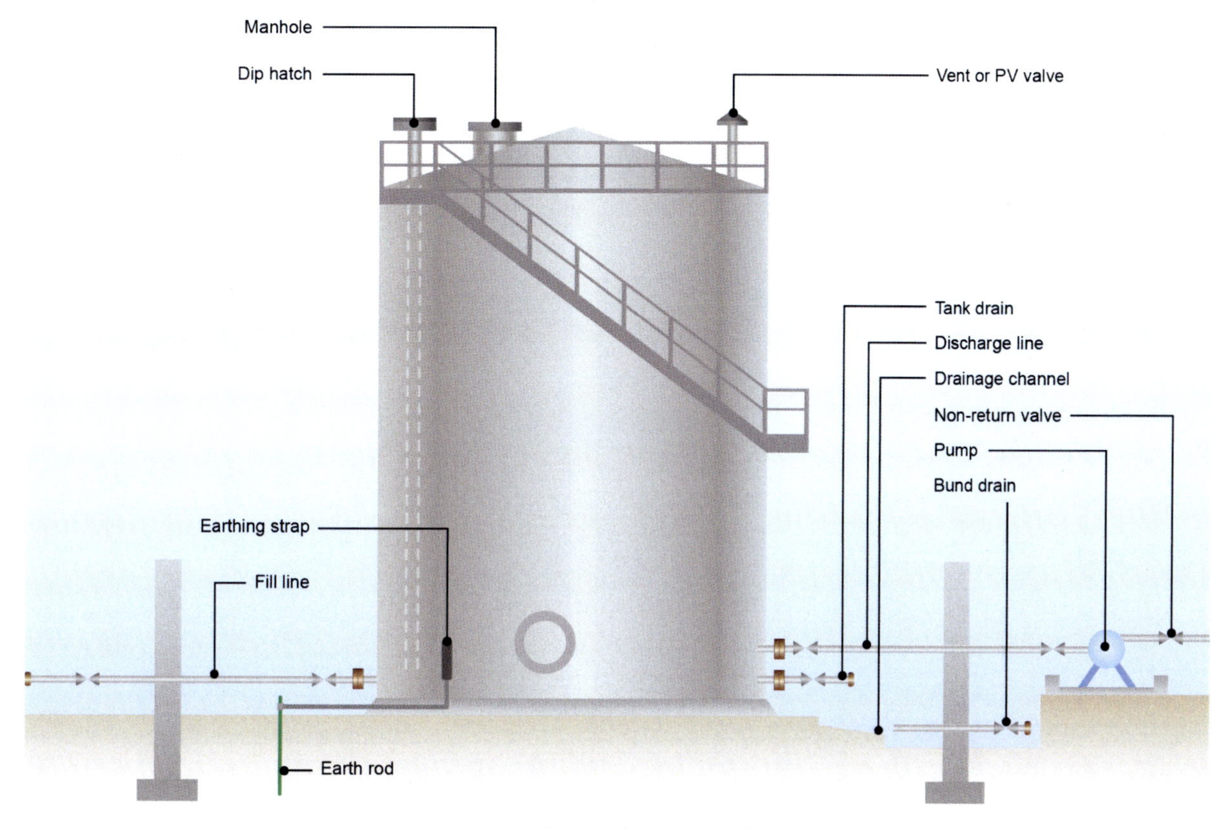

Typical vertical storage tank

The key control measures indicated in the above diagrams include:

- Venting of the storage tank to ensure that:
  - positive pressure doesn't build up during tank filling which might cause failure of the walls of the tank; and
  - negative pressure (partial vacuum) does not form in the tank during tank emptying which might cause the tank to collapse.
- Earthing of the tank and associated metalwork to ensure that static charge is not able to build up during filling and emptying operations, that might then lead to an electrical discharge.
- Siting of the tank inside a bund wall that is capable of retaining 110% of the contents of the tank in the event of loss of containment.

The siting of major installations should take account of the hazard presented to people beyond the site boundary. If the installation is subject to the **Control of Major Accident Hazards Regulations 2015 (COMAH)**, particular consideration should be given to the prevention and limitation of the effects of major accidents.

Recommended separation distances between tanks are set out in the HSE Guidance document HSG176 under 'Location and Layout of Tanks'.

MORE...

Information on the storage of flammable liquids in tanks can be found in HSE guidance document HSG176 at:

www.hse.gov.uk/pubns/priced/hsg176.pdf

Guidance on the safe storage of flammable liquids in containers up to 1,000L capacity at the workplace is available in HSG51 at:

www.hse.gov.uk/pubns/priced/hsg51.pdf

Information on the safe use and handling of flammable liquids in all general work activities is contained in HSG140 *Safe use and handling of flammable liquids*, available at:

www.hse.gov.uk/pubns/priced/hsg140.pdf

Intermediate Storage

Intermediate storage is used to refer to the use of storage containers that are capable of holding quantities of dangerous substances in the range of approximately 500 to 1,000L or 500 to 1,000Kg.

Examples include:

- **Intermediate Bulk Containers** (IBCs) for the storage of liquids - these are liquid containers inside a framework. They are either made from all metal, plastic in a metal frame or all plastic.
- **Flexible Intermediate Bulk Containers** (FIBCs or 'big bags', 'bulk bags' or 'dumpy bags') for the storage of solids such as granules or powders - these are usually made of plastic fabric.

IBCs and FIBCs can be stacked. Manufacturers' guidelines on maximum stack heights must be observed.

Drum Storage

Storage drums come in a range of sizes, shapes and materials of construction, but typical examples include:

- Steel drums with a capacity of 205L.
- Plastic drums with a capacity of 210L.

When drums are used for storage, there are a number of points to consider:

- Drums should be stored upright. If stored on their side, they must be properly chocked.
- They can be stacked upright (usually on pallets). Again, manufacturers' recommendations about maximum stack heights must be observed.
- They should be properly marked with signs showing the contents.
- Proper drum-handling equipment should be used - drums should not be rolled.
- They should be periodically examined for damage and removed from service if there are signs of damage.

Inadequate drum storage

One of the risks associated with plastic IBCs and plastic drums is that they are relatively easy to set alight, and once alight or exposed to fire the plastic may fail very rapidly, leading to a total loss of liquid contents. A relatively trivial fire, such as a grass fire, can ignite an IBC and cause loss of containment within a few minutes.

Intermediate bulk containers and drums should only be used for substances that they are designed for.

General Principles of Safe Storage in Specific Locations

Where possible, dangerous substances should be stored outdoors in an appropriately built and operated storage area. Where this is not possible or reasonably practicable, then an appropriately designed and operated indoor-storage facility must be used. The basic principles of safety for both intermediate containers and drums are very similar.

Outdoor Storage

Key design principles for **outdoor storage** of IBCs and drums are:

- Secure inside a locked compound to prevent unauthorised access, theft and arson.
- Clear signage on the compound to indicate the hazards of the contents and the key control measures such as no smoking and no naked flames.
- An impervious floor to the compound that is compatible with IBC/drum contents.
- A bund sill or wall to the compound capable of retaining 110% of the contents of the largest container.
- A slight slope to the floor of the bund so that leaks under IBCs/drums will flow out into open space where they will be visible.
- Any adjacent structures such as buildings must either have fire protection to the walls adjacent to the storage area or must be separated from the storage area by adequate distances.
- Any adjacent site boundaries must either be separated from the storage area by an adequate distance or fire protected with a fire-rated wall.
- Segregation and separation between incompatible substances (such as oxidising substances and incompatible chemicals; flammable liquids and oxidising substances, etc.).

Indoor Storage

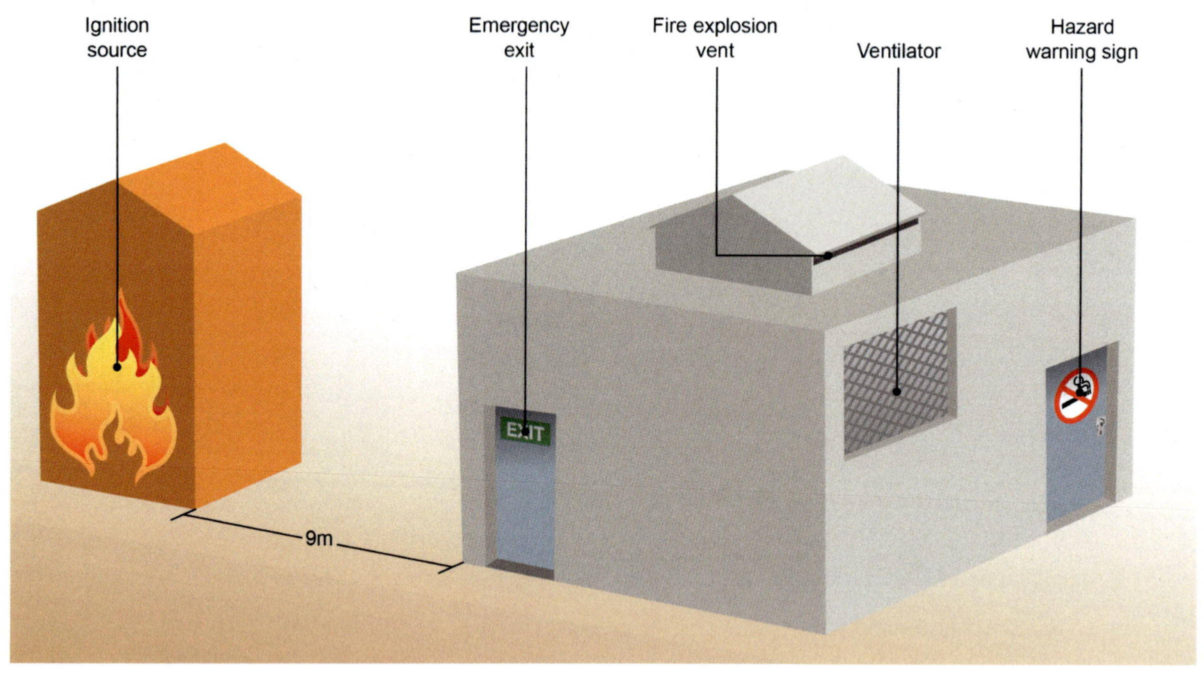

Dedicated flammable store showing typical safety characteristics and separation distance

Key design principles for **indoor storage** of IBCs and drums are:

- A dedicated storage building separated from other process or accommodation buildings where possible/ reasonably practicable.
- Where this is not possible, the storage area should be a fire-protected part of a single-storey building. Storage areas for dangerous substances should not form part of a multi-storey building unless very high levels of fire protection can be guaranteed and should not form part of any building that people use as sleeping accommodation.
- Security to prevent unauthorised access, theft and arson.
- Clear signage at entrances to indicate the hazards of the contents and the key control measures such as no smoking and no naked flames.
- An impervious floor to the room that is compatible with IBC/drum contents.
- A sill to the room to create a bund capable of retaining 110% of the contents of the largest container.
- A slight slope to the floor so that leaks under IBCs/drums will flow out into open space where they will be visible.
- Fire-rated walls to keep fire out of the storage area for a period of time.
- Self-closing, fire-rated doors to all entrances/exits.
- Fire-rated glass to any windows.
- A lightweight roof that will lift off easily in the event of a confined vapour cloud explosion in the building (so preventing overpressure from blowing out the walls).
- Good levels of natural ventilation through low-level and high-level vent/louvres.
- Segregation of incompatible materials.

In both cases the appropriate zone classification must be assigned and only electrical equipment that is safe for the relevant zone must be installed or taken into the zoned area.

Adequate fire-fighting equipment must also be provided in or adjacent to the area/room.

Safe Storage of LPG

'LPG' means either propane or butane or mixtures of the two. These gases are stored in steel cylinders under high pressure in a liquid form. So a full LPG cylinder consists of a steel cylinder with a valve assembly at its top, containing a liquid with a small vapour space at the top containing gas under high pressure.

LPG is, of course, a flammable gas and can cause fire and explosion if ignited. Large LPG cylinders such as 'torpedo' tanks used for bulk storage are pressure relieved (i.e. they have a pressure relief valve fitted).

For larger LPG tanks above ground, the HSE recommend separation distances from buildings, boundary property lines or a fixed source of ignition. In most cases:

- LPG tanks must be at least 3m away from buildings, boundaries and fixed sources of ignition.
- As LPG vapour is heavier than air, an LPG tank should not be positioned within 3m of unsealed drains or gullies.
- LPG tanks must be at least 1.5m away from overhead power cables.

MORE...

For a full guide on HSE-recommended separation distances, visit:

www.hse.gov.uk/gas/lpg/separationdistances.htm

Smaller storage cylinders (such as 47Kg upright cylinders used for domestic cooking/heating) are not pressure relieved. Both types of cylinder will cause a BLEVE if involved in a fire. With unrelieved tanks, this can occur in a very short period of time (minutes or seconds in extreme cases) due to the build-up of excessive pressure in a violent fire situation.

Smaller storage cylinders are particularly vulnerable to physical damage to the valve assembly. If a small cylinder is knocked over and its valve assembly strikes something, it may fail catastrophically leading to the rapid escape of gas and liquid under pressure which results in the cylinder taking off like a rocket.

Refer to the following figure for storage of LPG cylinders:

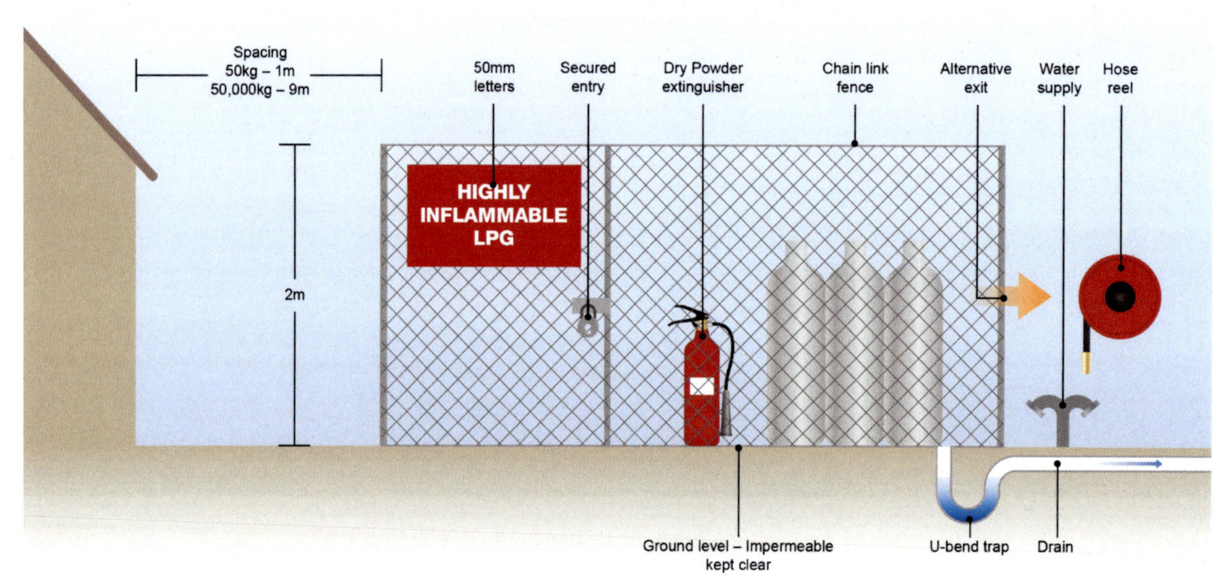

Storage of LPG cylinders

LPGs are to be stored in the following ways:

- LPG cylinders should be stored outside in a well ventilated area (e.g. not in the light-well of a building). It may be necessary to protect the storage area from direct sunlight exposure in some hot climates (not normally the case in temperate zones).
- LPG cylinders should be stored in a secure compound with a high fence to prevent unauthorised access.
- Cylinders should be stored upright and secured in the upright position (by chains) to prevent toppling.
- Cylinders should be segregated by content type.
- Full and empty cylinders should not be stored together.
- Acetylene may be stored in an LPG compound.
- Oxygen cylinders should NOT be stored in the same compound unless a significant separation distance can be maintained.
- The compound should be clearly signed to warn of the flammable gas hazard and to prohibit naked flames and ignition sources.
- The floor of the compound must be solid and impervious to LPG gas.
- Fire-fighting equipment (such as dry powder extinguishers or a hose reel) should be positioned by the compound to allow for easy fire-fighting of small fires inside the compound.

The storage area must be zoned (usually Zone 2) and only electrical equipment safe for that zone installed or taken inside.

Storage of Incompatible Materials

There are many materials that might be stored in a workplace that are potentially incompatible. They either significantly increase fire and explosion risk or can chemically react together to create significant chemical hazards.

Examples include:

- Oxidising substances (such as ammonium nitrate) and any form of fuel (such as flammable liquids or fuel oil).
- Acids and alkalis that will react together (perhaps explosively) to generate hazardous gases and vapours.
- Certain solids that will exothermically react with water to produce hazardous products and spontaneously ignite (e.g. calcium carbide and water react to form acetylene gas).

When dealing with such incompatible chemicals it is important to first recognise that there is an incompatibility issue. This requires careful assessment by a competent person, such as an industrial chemist, and enquiry into the reactive nature of the materials being stored by reference to information sources such as Safety Data Sheets (SDS).

Some of the key principles for safe storage of incompatible materials are:

- Store the materials indoors so that they can be protected from adverse weather and direct sunlight.
- Store in an appropriate building/part of a building with an adequate level of fire protection.
- Keep the chemicals secure at all times by ensuring that building entrances are kept locked and only authorised personnel gain entry.
- Use signage to communicate the hazardous nature of the stored chemicals and the rules for safe storage.
- Ensure that the floor of the building is compatible with, and impervious to, the stored chemicals.

Corrosive materials

- Ensure that an appropriate type of fire-detection system is installed in the building (e.g. heat, smoke or flame detection) and that appropriate fire-fighting media are available. The media must be compatible with the chemicals that are being stored.

Most importantly:

- Some chemicals, notably organic peroxides, must be **isolated** from, and therefore not stored in the same storage building as, other chemicals. These must be stored in a dedicated storage facility.
- Other chemicals such as oxidising substances and flammable substances must be **segregated** from one another. This means not stored in the same storage compound or within the same building fire compartment. Compartments within a building should have at least 30 minutes' fire-resistance between segregated chemicals.
- Lower-risk incompatible chemicals, such as flammable substances and corrosives can simply be **kept apart**. This means that they are stored within the same compound or building compartment, but with a minimum separation distance of three metres.

Toxic substances

It is important to recognise that incompatible substances must not be stored above one another, for obvious reasons.

Leakage and Spillage Containment

Means of **controlling spillage** should be provided, including absorbent granules or other means of clearing up small spills, where appropriate. This may be an impervious sill or low bund, typically 150mm high, big enough to hold 110% of the contents of the largest container. An alternative is to drain the area to a safe place, such as a remote sump or a separator.

Bunding

A bund is an enclosure around a storage facility designed so that in the event of any leak or spillage from tanks or pipe work, it will capture well in excess of the volume of liquids held within the bund area. It is usually designed to hold at least 110% of the contents of the liquid storage vessels, tanks or drums that it surrounds without their being able to escape. Bunds are usually made of concrete or masonry but can be metal.

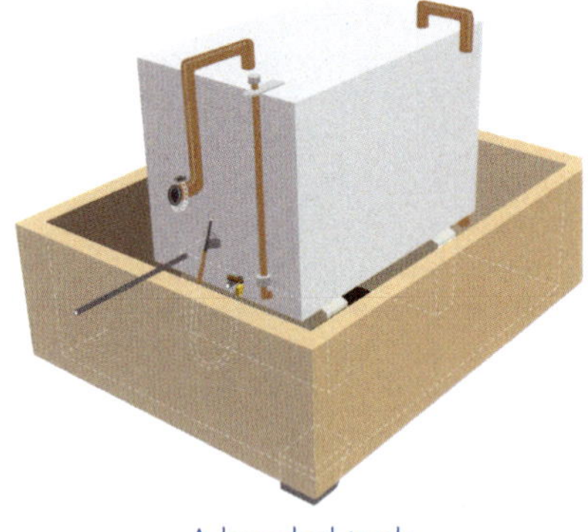

A bunded tank

Bunds may contain more than one tank and should be designed to hold at least 110% of the capacity of the largest tank within the bund, after making allowance for the space occupied by other tanks. In exceptional cases, where there is no risk of pollution or of hazard to the public, this figure may be reduced to 75%.

Intermediate lower bunds may be used to divide tanks into groups to contain small spillages and to minimise the surface area of any spillages. This will reduce the maximum size of a bund fire.

Bunds should be substantially impervious to the liquid being stored and designed to withstand the full hydrostatic head. They may be partly below ground level to help provide adequate wall strength. Impact protection, such as crash barriers and bollards, should be provided where necessary.

The height of the bund should take account of the need to ensure adequate ventilation within the bund, ready access for fire-fighting and good means of escape. The height should not normally exceed 1.5m, although 2m can be used in some cases for large tanks. Diversion walls and intermediate lower bunds should not exceed 0.5m unless there are special circumstances, such as sloping ground.

The floor of the bund should be of concrete or other material which is impervious to the liquid being stored, and with drainage, where necessary, to prevent minor spillages collecting near tanks. Stone chippings and similar materials may be used, providing the underlying ground is impervious. A suitable buried membrane can also be used, as can specially designed systems using the water table to retain liquids not miscible with water. No vegetation (except short grass) or other combustible material, and no liquid containers or gas cylinders (full or empty) should be stored in the bund or within 1m of the outside of the bund wall. Weedkiller containing sodium chlorate or other oxidising substances should not be used at storage areas or tanker stands because of the increased fire hazard.

The surface of the storage area should be sloped so that any spillage does not accumulate around containers but can drain to an evaporation area (within the storage area or separate from it) or to a sump or interceptor.

Means of removing surface water from the bund should be provided. If an electrically-driven pump is used, the electrical equipment should be of a type suitable for the zone in which it is used.

If a bund drain is used, it should have a valve outside the bund wall, with a system of work in force to ensure the valve remains closed, and preferably locked, except when water is being removed. Surface water from bunds where flammable liquids not miscible with water are stored should be routed through an interceptor to prevent flammable liquids entering the main drainage system. For liquids miscible with water, special drainage systems may be required.

Many small storage tanks achieve the bund by constructing one tank inside another. This arrangement provides catchment in the event of failure of the storage tank and eliminates the problem of the bund area filling with rainwater.

Problems Encountered During Filling and Transfer

Filling and transfer of flammable liquids should be carried out in such a way that spills and dangerous releases of flammable vapours are reduced or eliminated if possible.

Problems involving flammable liquids commonly arise during filling and transfer operations, including:

- movement from storage and within premises;
- decanting and dispensing;
- process activities;
- emptying plant and equipment, including vehicle fuel tanks prior to maintenance;
- disposal;
- dealing with spillages.

In the first instance, an enclosed transfer system involving the use of pipeworks, pumps and closed vessels should be used. If the system is pressurised then care should be taken to ensure an inert gas, such as nitrogen, is used, as compressed air can create an explosive vapour/air mixture inside the transfer vessel.

The gas pressure should be kept as low as is practically possible with the facility for rapid isolation of the pressure source and venting of the pipeline to a safe place. Transfer into empty vessels by pressure feed is not advised as it is difficult to control the flow adequately and there is a risk of pressurising the container, causing it to rupture violently.

Poorly maintained transfer equipment (pumps, lines and valves) will leak and may even fail, leading to a major leak of flammable liquid, fires and explosions.

MORE...

The HSE website has a case study on the 1996 fire at Albright and Wilson, Avonmouth which you can access here:

www.hse.gov.uk/comah/sragtech/casealbright96.htm

Storage and Handling of Dangerous Substances

One of the main dangers associated with the use and storage of flammable liquids is the release of vapours. This must be prevented wherever possible, or at the very least kept to an absolute minimum.

In storage, vapour escape can occur when a substance is decanted into a smaller receptacle for use, allowing vapour to escape. Decanting in a storage area should be avoided.

The preferred method of transfer is by an enclosed pipe network to the point of use. This reduces the possibility of vapour escape and accidental spillage.

A simple filling operation with petrol

CASE STUDIES

Texas City (USA) 2005

Fifteen people (mainly contract workers) were killed and 180 others were injured in an explosion and subsequent fire that occurred at BP's refinery in Texas City, Texas, USA. A vivid example of the consequences of loss of containment resulting in an uncontrolled release of highly flammable liquid.

On 23 March 2005, a series of explosions occurred during the re-starting of a hydrocarbon isomerisation unit at BP's refinery in Texas City. A distillation tower flooded with highly flammable liquid hydrocarbons and was over-pressurised, causing a release of liquid from the top of the stack and a cloud of flammable vapour to form over the refinery. A diesel pick-up truck that was idling nearby ignited the vapour, initiating a series of explosions and fires that swept through the unit and the surrounding area.

The US Chemical Safety and Hazard Investigation Board investigating the incident found that approximately 7,600 gallons of hot, flammable liquid hydrocarbons - nearly the equivalent of a full tanker truck of gasoline - were released from the top of the blow-down drum stack in just under two minutes. The ejected liquid rapidly vapourised due to evaporation, wind dispersion, and contact with the surface of nearby equipment. High over-pressures from the resulting unconfined vapour cloud explosion totally destroyed 13 trailers and damaged 27 others. People inside trailers as far as 500 feet away from the blow-down drum were injured, and trailers nearly 1,000 feet away sustained damage.

The distillation tower overfilled because a valve allowing liquid to drain from the bottom of the tower into storage tanks was left closed for over three hours during the start-up on the morning of 23 March, which was contrary to unit start-up procedures. The investigation found that procedural deviations, abnormally high liquid levels and pressures, and dramatic swings in tower liquid level were the norm in almost all previous start-ups of the unit since 2000. Operators typically started up the unit with a high liquid level inside and left the drain valve in manual - not automatic - mode to prevent possible loss of liquid flow and resulting damage to a furnace that was connected to the tower. These procedural deviations - together with the faulty condition of valves, gauges, and instruments on the tower - made the tower susceptible to overfilling.

Buncefield (UK) 2005

Another example of the serious consequences of loss of containment of flammable liquids is the Buncefield incident, a major fire caused by a series of explosions in the early hours of Sunday 11 December 2005 at the Buncefield Oil Storage Depot, Hemel Hempstead, Hertfordshire. At least one of the initial explosions was of massive proportions and there was a large fire which engulfed a high proportion of the site. Over 40 people were injured but there were no fatalities. Significant damage occurred to both commercial and residential properties in the vicinity and a large area around the site was evacuated on emergency service advice. The fire burned for several days, destroying most of the site and emitting large clouds of black smoke into the atmosphere.

The cause of the incident was the overfilling of a storage tank, allowing fuel to cascade down the side of the tank leading to the rapid formation of a vapour cloud that subsequently ignited, leading to the explosions and fire. The sequence of events was as follows:

- 10 December 2005: Around 19:00, Tank 912 in bund A at the Hertfordshire Oil Storage Ltd (HOSL) West site started receiving unleaded motor fuel from the T/K South pipeline, pumping at about 550m^3/hour.
- 11 December 2005: At approximately midnight, the terminal was closed to tankers and a stock check of products was carried out. When this was completed at around 01:30, no abnormalities were reported.

(Continued)

CASE STUDIES

- From approximately 03:00, the level gauge for Tank 912 recorded an unchanged reading. However, filling of Tank 912 continued at a rate of around 550m^3/hour. Calculations show that at around 05:20, Tank 912 would have been completely full and starting to overflow. Evidence suggests that the protection system, which should have automatically closed valves to prevent any more filling, did not operate.
- From 05:20 onwards, continued pumping caused fuel to cascade down the side of the tank and through the air, leading to the rapid formation of a rich fuel/air mixture that collected in bund A.
  At 05:38, CCTV footage shows vapour from escaped fuel starting to flow out of the north-west corner of bund A towards the west. The vapour cloud was about 1m deep.
- At 05:46, the vapour cloud had thickened to about 2m deep and was flowing out of bund A in all directions.
- Between 05:50 and 06:00, the pumping rate down the T/K South pipeline to Tank 912 gradually rose to around 890m^3/hour.
- By 05:50, the vapour cloud had started flowing off-site near the junction of Cherry Tree Lane and Buncefield Lane, following the ground topography. It spread west into Northgate House and Fuji car parks and towards Catherine House.
- At 06:01 the first explosion occurred, followed by further explosions and a large fire that engulfed over 20 large storage tanks. The main explosion event was centred on the car parks between the HOSL West site and the Fuji and Northgate buildings. The exact ignition points are not certain, but are likely to have been a generator house in the Northgate car park and the pump house on the HOSL West site.
- At the time of ignition, the vapour cloud extended to the west almost as far as Boundary Way in the gaps between the 3-Com, Northgate and Fuji buildings; to the north-west it extended as far as the nearest corner of Catherine House. It may have extended to the north of the HOSL site as far as British Pipelines Agency (BPA) Tank 12 and may have extended south across part of the HOSL site, but not as far as the tanker filling gantry. To the east it reached the BPA site.

There is evidence suggesting that a high-level switch, which should have detected that the tank was full and shut off the supply, failed to operate. The switch failure should have triggered an alarm, but it too appears to have failed.

The Health Protection Agency and the Major Incident Investigation Board provided advice to prevent incidents such as these in the future. The primary need was for safety measures to be in place to prevent fuel from exiting the tanks in which it is stored. Added safety measures were needed for when fuel does escape, mainly to prevent it forming a flammable vapour and stop pollutants from poisoning the environment.

Flow Through Pipelines

It is often necessary to move and deliver gases, liquids or solids (powders, etc.) to a particular point as part of a process. One common method of transporting these substances, including dangerous substances, and ensuring that they are isolated from workers, is to send them through enclosed pipelines to the point of delivery or use. Not only is it a convenient method of transporting substances, but the chance of escape and subsequent exposure is reduced when compared with other methods such as containers and drums with opening lids.

Pipeline transfer

The oil industry is one typical example of a process that uses pipelines; the image of a lone pipeline crossing a desert and disappearing into the far horizon is common. At the other end of the spectrum, another example of pipeline use is a hosepipe. While the difference in scale of the two examples is vast, the purpose is the same - to deliver a substance to a desired point. A further example of the use of pipelines in industry is for the purpose of unloading road tankers. Typical substances that may be carried include: petrol, gases, and solids such as flour.

There are a number of issues that need to be considered, however, when using pipelines to transport substances. Depending on the length of the pipeline, the diameter of the pipe and the substance flowing through the pipe, pressure will have to be applied to ensure a satisfactory flow. While this pressure may not always be particularly high, it can cause problems if there is a weakness in the pipeline, particularly at joints and flanges. Any weakness can result in the failure of the integrity of the pipeline, resulting in a release of the substance and a reduction in flow. This may have hazardous results for the process, with a possible lack of feedstock or other important components. Similar problems can arise from a blockage in the pipeline causing overpressure.

The effects of blockages can be easily demonstrated by standing on a hosepipe and observing the weak points of the system where water leaks out. If there are no weak points and water cannot escape, then the system will eventually fail due to the pressure build-up. Many pipelines have built-in pressure relief systems to deal with these potential problems.

Another aspect of pressure is that it affects the flow-rate of the substance. The flow-rate, in turn, is an important factor in the generation of electrostatic charges or 'static electricity'. This is caused by the movement of molecules of dissimilar materials against each other. As the two materials come into contact and then separate, electrons are exchanged and this causes a charge imbalance between the dissimilar materials. When the surfaces are separated, some bonded atoms are left with extra electrons (negative charge) and some with a deficit of electrons (positive charge). This build-up of charge on insulating materials creates a difference in electrical potential which, if high enough, can generate a spark.

For example, petrol flowing through a plastic pipe will generate an electrostatic charge, as will a powder such as flour running down a steel chute.

The main danger associated with an electrostatic charge is the possibility of a spark occurring if the charge is not allowed to dissipate to earth. If the spark occurs in an explosive environment, and exceeds the minimum ignition energy of the explosive mixture, the results can be catastrophic. For example, if all elements of the pipeline, including discharge nozzles, are not equipotentially bonded and earthed, then the charge is quite capable of 'jumping' a gap and producing a spark. The blame for a number of accidents at filling stations has been apportioned to a static spark between the nozzle of the petrol pump hose and the filler on the car.

When dealing with any type of substance flow through pipelines, it is important to be aware of the possibility of a spark from an electrostatic charge, regardless of the amount of substance and distance travelled. Bonding and earthing of all parts of a pipeline system is vital to ensure that any charge safely escapes to earth.

Principles in Filling and Emptying Containers

Filling Containers

There are a number of issues to consider when filling containers. Perhaps the most obvious is to ensure that the container that is being used is **suitable** for the substance that is being put into it. Whether it is a small 'emergency' petrol container or a large acid tank or drum, it is important to ensure that it is designed to accept the substance. It must also be **undamaged** with no signs of leakage, staining or corrosion.

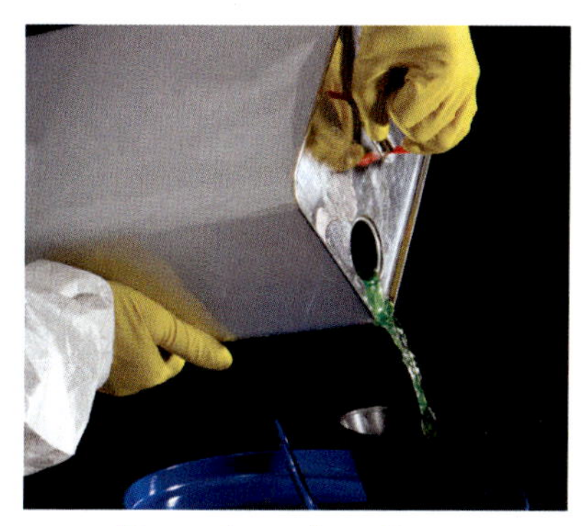

Dispensing solvent into a storage drum

Once satisfied that the container is suitable, the next thing is to determine whether it is **empty and clean**. If it is not, then the previous content must be established. Mixing incompatible substances can lead to uncontrolled chemical reactions. When dealing with volatile substances it may be necessary to have the container de-gassed and certificated 'clean' prior to filling.

The method of filling has safety implications. The most common methods are:

- **'Top' filling** which, as the term suggests, is achieved by the use of a hose or tap through an opening in the top of the container (similar to old steam engines being watered). This method of filling is often used when decanting into small containers and drums.
- **'Bottom' filling**, where the substance is delivered into the container under pressure through a closed pipeline. This is a common method of filling road tankers.

Both systems have a degree of risk associated with them:

- Top filling can create 'splash', which can contaminate the surrounding area as well as generating a large electrostatic charge. It will also allow the escape of vapours.
- Bottom filling through a closed system alleviates the problem of vapour escape, but the pressure under which the substance is delivered can cause problems unless the container is designed to be pressurised and any pressure venting and release devices are functioning properly.

Regardless of the method of filling, care has to be taken to avoid **overfilling**. The consequences of overfilling are potentially enormous with uncontrolled contamination of the environment and release of vapours. Care should be taken to ensure that the receiving container is large enough to receive the amount of substance that is going to be delivered and that the process is supervised during the delivery process. Bunding and spill control kits should be available in case of a spillage.

The **proper marking** of suitable containers is also important as unmarked or incorrectly marked containers can lead to tragic accidents.

Emptying Containers

Factors such as spillage, escape of vapour and generation of electrostatic charges are just as relevant when emptying containers. The flow of substance from even a very small container can generate enough electrostatic charge to develop a spark large enough to cause an explosion under the right conditions.

Other factors to consider are:

- **Nominally Empty Containers**

  These are containers that have been emptied of the substance but may still contain vapour. Containers in this condition are potentially very dangerous and should be closely stoppered until cleaned or disposed of safely. The practice of leaving nominally empty petrol containers open to 'breathe' should be discouraged as the vapour, being heavier than air, will remain in the container and mix with air, making an explosive mixture.

- **Implosion of Container**

  This can happen as a result of, for example, a liquid being discharged out of the bottom of an otherwise sealed container/tank. If account is not taken of the need to replace the discharging substance with air/vapour (or another medium in certain circumstances), then a partial vacuum can develop and cause the sides of the container to implode. One way of avoiding this is, of course, to vent the top of the container (e.g. by opening a hatch on the top of a bottom-discharge tanker). Sometimes, though, it is desirable or necessary to use completely closed systems to stop the vapour from escaping. This may be desirable for a variety of reasons, not least, environmental protection. In such cases, instead of venting to the atmosphere, a vapour return line is connected between the discharging vessel and the receiving vessel (see following figure). In this way, the liquid being discharged is replaced with the vapour displaced from the top of the receiving vessel. This is the case with road tankers transferring petrol. The vapour return line may be separate from the liquid transfer line or be incorporated within the transfer line.

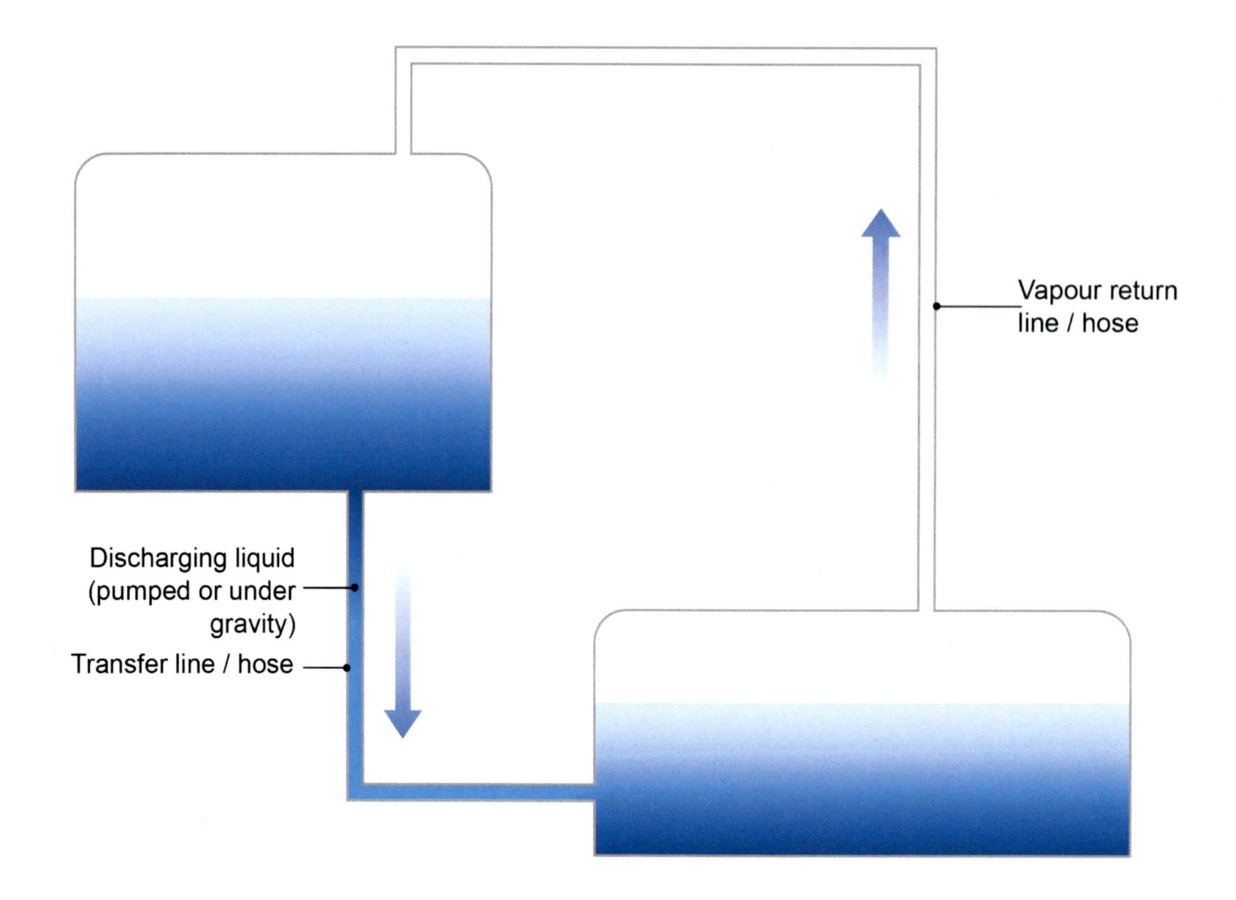

Transfer of liquid using vapour return valve

- **Steam Cleaning**

  Implosion can also happen after cleaning if steam has been used. If the container does not allow air in as the steam cools and condenses (remember that steam is water expanded 1700 times and when it turns back into water there is a consequent reduction in volume), this reduction of volume can cause an implosion of the vessel. Hatches on tanks that have been steam-cleaned should be left open until the temperature inside the tank has dropped to a point where implosion will not occur.

Reduction of Risks

In order to reduce the risks associated with the filling and emptying of containers, certain measures, in addition to those already mentioned, should be taken:

- A risk assessment should be carried out.
- All ignition sources should be controlled.
- All personnel involved with the process should be properly trained and supervised.
- Empty and full containers should be segregated.
- All containers should be examined for damage.
- Suitable signs and notices should be displayed.
- Suitable fire-fighting equipment should be available.

MORE...

The HSE publication L133 *Unloading petrol from road tankers - Dangerous Substances and Explosive Atmospheres Regulations 2002 - Approved Code of Practice and guidance* provides guidance for persons involved in the delivery and unloading of petrol at filling stations, such as petrol station site operators, road tanker operators, road tanker drivers and authorities who have responsibility for enforcement of **DSEAR 2002** at petrol filling stations.

The second edition is available at:

www.hse.gov.uk/pubns/books/l133.htm

Principles in Dispensing, Spraying and Disposal of Flammable Liquids

Dispensing or decanting should generally be avoided. Large volumes are best transferred in closed systems, such as pipelines. Small volumes may be dispensed, but this activity must not be carried out in areas designed for storage of dangerous substances. Such an activity may create a risk of fire involving the stored materials.

Precautions should be taken against:

- Accidental spillage during dispensing, e.g. by use of spill trays or other bunding-type arrangements to limit liquid spread.
- Excessive flammable vapour release by only using tight-head vessels (i.e. narrow orifice containers) that, preferably, are fitted with flame arresters and design features to minimise the build-up of electrostatic charge.

Spray-painting furniture with flammable varnish on an assembly line

Dispensing should be carried out in a well-ventilated area in order to quickly dilute any escaping vapour concentrations to below the LEL and should take place away from any ignition source, to prevent fires or explosion. This could include:

- Control of static by using, for example, earthed metal containers, and bonding clips prior to decanting.
- Prohibition of smoking.
- Use of non-sparking tools.

Empty, uncleaned containers should be treated as if they are full, as they will contain flammable residues.

Spraying of flammable liquids, such as paints, may easily create a flammable atmosphere - especially when conducted in a confined area with poor ventilation. The vapour concentration may rapidly enter the explosive range. It is therefore essential to control both the vapours and ignition sources during such operations, otherwise the vapour may be ignited, causing a fire or explosion. Usually, spraying operations are conducted in well-ventilated spray booths. The ventilation should be sufficient to dilute the flammable vapours to well below the LEL.

The spray booth should be isolated from all ignition sources. Spray booths are typically classified as Zone 1 areas under Regulation 7, Schedule 2 of the **DSEAR** - where an explosive gas/air mixture is likely to be present during normal operation. Any electrical equipment in the booth should therefore take this into account. Liquid movement during spraying, and during filling, emptying, etc. can cause static build-up. Ignition by static discharge is a particular problem, and precautions should be taken against the build-up of static - use of anti-static footwear, clothing and flooring, and the earthing of all equipment.

Waste should be stored appropriately prior to **disposal** - in suitable closed containers that are compatible with the contents. Wastes having different compositions should not be mixed together without regard to possible dangerous reactions. Due account should be taken of the applicable environmental legislation controlling the disposal of waste.

TOPIC FOCUS

The measures that should be taken to control the fire and explosion risks associated with **spray-painting a solvent-based paint with a low flash point** include:

- Fire-resistant workplace construction.
- Use of a less flammable, solvent-based substitute paint.
- Reduce the quantity of paints and solvents to a minimum.
- Provide an external, fire-resistant storeroom for paints and solvents with adequate ventilation, located an adequate distance from the workplace.
- Provide local exhaust ventilation.
- Provide dilution ventilation at high and low levels.
- Use intrinsically safe electrical equipment.
- Earth to avoid electrostatic ignition.
- Use conductive footwear and clothing to avoid static build-up.
- Identify the appropriate zone for the work area.
- Use non-spill containers.
- Provide fire-fighting equipment, escape routes and an emergency plan.

Dangers of Electricity in Hazardous Areas

Electricity, and its associated uses, are acknowledged sources of ignition in the workplace and, as such, need to be closely controlled. This becomes even more compelling when electricity is used in areas in which the environment is, or has the capability to be, flammable or explosive. This atmosphere may be caused by vapours from flammable substances, flammable gas or even the presence of dust. Under these circumstances, all of the elements of the fire triangle are present, and fire or explosion would seem to be highly probable. As we saw earlier in the unit, **DSEAR** identifies 'zones' which relate to the presence, or possible presence, of flammable atmospheres and the **Equipment and Protective Systems Intended for Use in Potentially Explosive Atmospheres Regulations 2016** legislate on suitable electrical equipment that may be used in those zones.

Transport of Dangerous Substances

Carrying goods by road or rail involves the risk of traffic accidents. If the goods carried are dangerous, there is also the risk of an incident, such as spillage of the goods, leading to hazards such as fire, explosion, chemical burn or environmental damage. Most goods are not considered sufficiently dangerous to require special precautions during carriage. Some goods, however, have properties which mean they are potentially dangerous if carried.

Dangerous goods are liquid or solid substances and articles containing them that have been tested and assessed against internationally-agreed criteria (a process called classification) and found to be potentially dangerous (hazardous) when carried.

Dangerous goods are assigned the following different classes depending on their predominant hazard:

1. Explosive substances and articles.

2. Gases.

3. Flammable liquids.

4.1 Flammable solids, self-reactive substances and solid desensitised explosives.

4.2 Substances liable to spontaneous combustion.

4.3 Substances which, in contact with water, emit flammable gases.

5.1 Oxidising substances.

5.2 Organic peroxides.

6.1 Toxic substances.

6.2 Infectious substances.

7. Radioactive material.

8. Corrosive substances.

9. Miscellaneous dangerous substances and articles.

There are regulations to deal with the carriage of dangerous goods which are discussed briefly below. Their purpose is to protect everyone either directly involved (such as consignors or carriers), or who might become involved (such as members of the emergency services and public). Regulations place duties on everyone involved in the carriage of dangerous goods to ensure that they know what they have to do to minimise the risk of incidents and guarantee an effective response.

Carriage of dangerous goods by road or rail is regulated internationally by agreements and European Directives, with biennial updates of the Directives to take account of technological advances. New safety requirements are implemented by EU member states via domestic regulations which, for Great Britain, directly reference the technical agreements.

Key Safety Principles in Loading and Unloading of Tankers and Tank Containers

Statistical analysis of accidents indicates that 80% of transport-related incidents and accidents occur during loading and unloading operations. Further detailed analysis shows that in 90% of these cases, the human factor is the root cause.

Road tankers carrying petroleum spirit or other substances are sometimes referred to as 'fixed tanks' or road tank vehicles; the tank is permanently fixed to the vehicle chassis. Tank containers (sometimes called 'ISO tanks' or 'portable tanks') are held in a boxed-steel framework. The framework is locked to the vehicle chassis but can be unloaded from the vehicle - this is particularly suited to transfer of tanks between, say, road vehicle and train, or road vehicle and ship.

A petrol tanker

Safety principles in loading and unloading of tankers and tank containers include:

- Drivers of tankers should check in at the gate security/administration. At this point, the load should be checked against the load manifest by the on-site security/administration (see: 'ADR Instructions in Writing' below) and the driver given site safety and behaviour instructions and directions to the loading bay.
- Whilst on site, drivers must wear the correct PPE including a safety hat, safety shoes, high-visibility vest, gloves (EN 374 JKL or according to product) and working clothes that fully cover the arms and legs at all times.
- If loading/unloading chemicals then protective clothing specific to that chemical should be worn (protective clothing with limited flame spread properties with protection against the danger caused by static electricity and protection against liquid chemicals, type EN 533, EN 1149/5, or EN 13034 type 6).
- Ear protection and safety goggles should be worn where indicated and if the driver is required to access the top of the tank, then a safety harness/fall protection should be used.
- All other personnel should be excluded from the loading/unloading area.
- In the event of an on-suite alarm then drivers should follow instructions provided by the site management.
- After entering the site, drivers should proceed directly to the loading location.
- There must be free access and egress to and from the loading bay. If the driver has inadequate visibility for manoeuvring, a banksman must be provided by the site to ensure the safety of the vehicle and others using the area.
- Adequate signage should be provided by the site to indicate one way roads, pedestrian crossings and speed limits.
- At the loading location, drivers should switch off the engine, and engage the handbrake.

Filling and emptying road tankers involves the risk of fire and explosion if a flammable mixture of fuel and air is generated above the explosive limits in the presence of an ignition source. Consequently, control measures need to be in place to:

- Prevent the formation of a flammable mixture of fuel and air:
  - Vapour return systems to reduce flammable vapour release.
  - Adequate tank space to prevent spillage through overfilling.
  - Level monitoring with alarms.
  - Nitrogen blanketing of the road tanker and the bulk storage tank.
  - Monitoring equipment to detect leaks from the tank and associated (buried) pipework.
  - Strict operating procedures to prevent leaks and spills.
- Control potential ignition sources:
  - Prohibition of smoking.
  - Zoning of the filling area and electrical equipment appropriate for that zone.
- Reduce the risk of ignition by static electricity by:
  - Control of pumping rates.
  - Proper pipe sizing to keep liquid velocities low.
- Elimination of splash filling and free-fall of flammable liquids by:
  - Lowering fill velocities.
  - Directing the discharge of liquid down the side of the vessel.
  - Submerging fill pipes below the liquid level in the vessel.
  - Avoiding the use of filters or installing filters far enough upstream of discharge points to allow adequate time for any static generated to leak away.
  - Use of antistatic footwear and clothing.
  - Earthing of pipeline, vehicle and tank.
  - Electrical bonding of all pipe joints and of the pipeline to the tanker.
- Deal with emergencies:
  - Have fire extinguishers available to deal with any small outbreaks of fire on, or in the immediate vicinity of, the fuelling unit.
  - Dry sand or other absorbent material to aid the clearing up of small leaks or spills.

Hose Couplings

Several (non-compatible) couplings are commonly used in Europe for the (un)loading of chemicals and gases and there are even more product-specific types used in dedicated supply chains.

This is a significant disadvantage for the flexibility of the European supply chain and has proven to be a substantial obstacle for loaders, transporters and discharge locations alike.

Every supply chain solution is optimised around its chosen coupling and, as a result, both sites and transport companies have a large collection of adaptors to go from one type to the other.

This practice is bad for quality (as those adaptors and their respective gaskets are not cleaned/replaced on every occasion), bad for safety (as the sequencing of connection pieces exponentially increases the risk of spills) and bad for the environment as the sheer number of connection pieces required increases the weight of the transport unit which, in turn, has an unnecessary CO_2 cost for the transport.

Several attempts to standardise coupling have failed since each one has its own specific advantages and disadvantages but no consensus has been reached on this subject so far.

ADR Instructions in Writing

Drivers of tankers must be fully informed of dangers of the materials carried and the emergency action that needs to be taken.

The ADR Instructions in Writing have replaced the previous Tremcard system and are now only required in the language of the driver (or crew) for the journey.

These instructions must be provided by the carrier to the vehicle crew in language(s) that each member can read and understand before the commencement of the journey. The carrier should ensure that each member of the vehicle crew concerned understands and is capable of carrying out the instructions properly.

Before the start of the journey, the members of the vehicle crew must inform themselves of the dangerous goods loaded and consult the Instructions in Writing for details on actions to be taken in the event of an accident or emergency.

The Instructions in Writing must be kept in the vehicle cab so that they can be easily located by the emergency services in the event of an accident. The driver and the recipient of materials should have written procedures that set out the precautions that need to be taken during loading and unloading. Fire extinguishers should be carried on all vehicles. If substances are flammable or explosive, earth connections should be used during loading and unloading to prevent the possibility of a static spark, and no other sources of ignition, such as smoking materials, should be allowed in the vicinity.

Where bulk storage tanks are used for different substances, there is always the possibility of cross-contamination - a substance being unloaded from a tanker into the wrong bulk tank at a factory. This can be prevented by strict operating procedures and the use of couplings of a different design for each substance. It is also important to ensure that tanks to be filled have enough space so as to prevent spillage through overfilling.

Labelling of Vehicles and Packaging of Substances

As part of the regulatory system that controls transportation of dangerous goods in the UK, there is a requirement that vehicles are **placarded** and **marked**.

Placarding is the display of hazard warning information, which may contain graphics communicating the hazardous nature of the load for the benefit of the emergency services attending an accident involving the vehicle.

Marking is the display of orange reflectorised rectangular plates that may provide information on:

- Emergency action (the emergency action code).
- The material being transported (UN number).
- Address and telephone number of the consignor.

For UK national transport the 'HazChem' panel is used, shown below:

'HazChem' panel

- The emergency action code (or 'HazChem' code) gives advice to the fire authority on action to take.
- The United Nations (UN) number is specific to the substance (or type of substance) being carried.
- The box in the lower right-hand corner identifies the company consigning the load.
- The telephone number is used for contact purposes in an emergency.

When dangerous goods are transported in packages (rather than in tankers), the package needs to be labelled and marked indicating the hazardous nature of the contents:

- The hazard warning symbol.
- The UN number.

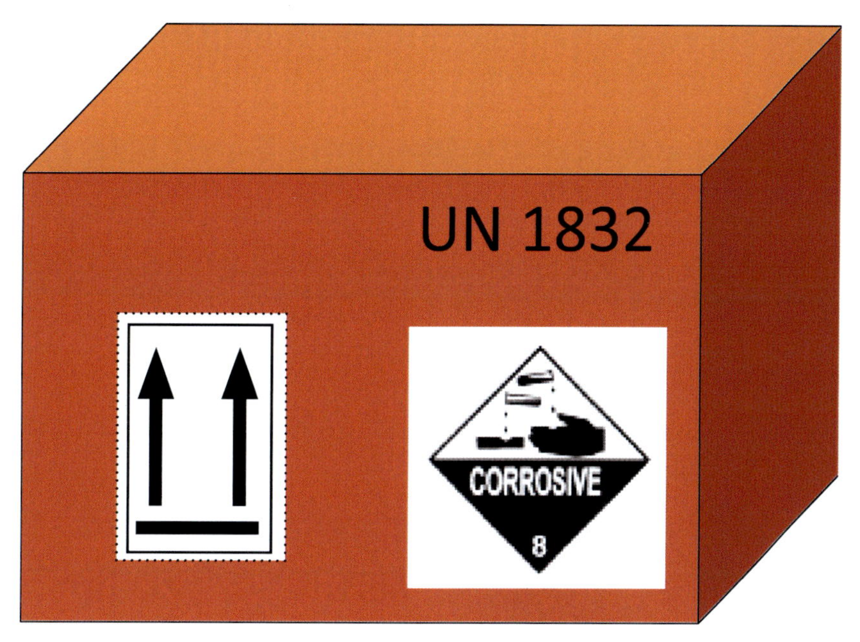

Example of a UN type-approved design

The packaging itself should be of a UN type-approved design:

- Tested and certified and marked as such. Tests include:
  - Stacking.
  - Impact.
  - Internal pressurisation.
- Every package, having passed the UN approval procedure, may carry the 'UN' approved symbol (a 'u' over an 'n' within a circle); this symbol is followed by a code which indicates the standard type of packaging, its performance and approving body.

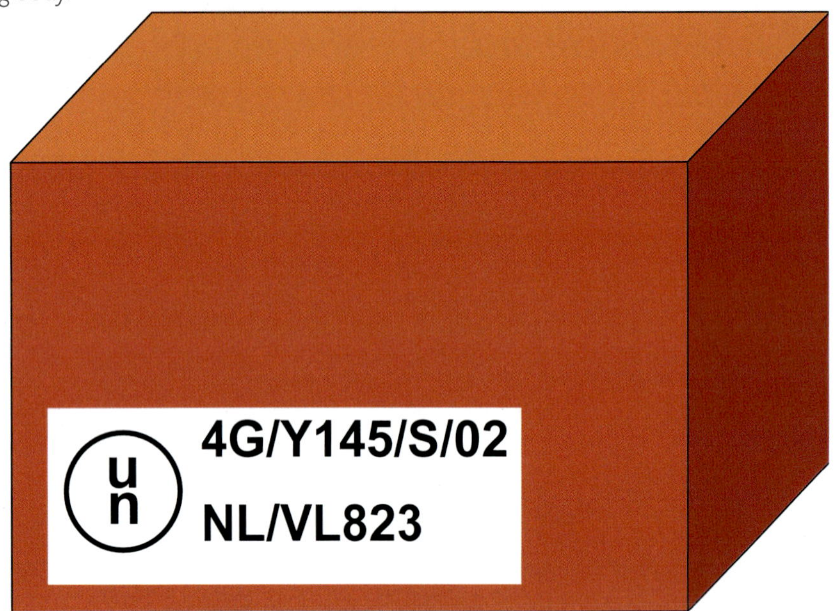

Example of a UN approval symbol

Driver Training and the Role of the Dangerous Goods Safety Adviser

The transport of dangerous substances is a highly complex and specialised area. Consequently, it is sufficient for the purposes of this course simply to be familiar with the broad regulatory framework and the key requirements to minimise the associated risks. Transport of dangerous substances may involve transport across national boundaries and therefore is heavily regulated by legally enforced international agreements such as:

- ADR - for European road transport.
- IMDG - for sea transport.
- ADN - for European inland waterways.
- RID - for European rail transport.
- ICAO/IATA - for air transport.

There are also special rules for the Channel Tunnel.

The current UK regulatory system references an international agreement known as 'ADR' (the letters taken as an abbreviation of the French title for the document). ADR is a very extensive and detailed document covering all the conceptual and operational requirements for European road transport of dangerous goods. It is normally updated on a two-year cycle. The UK has certain exemptions, extensions and modifications but the bulk of ADR is enforced 'as is'.

Road transport within the UK is mainly controlled under the **Carriage of Dangerous Goods and Use of Transportable Pressure Equipment Regulations 2009 (Carriage Regulations)**. These Regulations reference ADR but have some UK-specific modifications. Enforcement is mainly the responsibility of the HSE but some responsibility may fall on the Department for Transport, the Office of Rail and Road or the police.

The Regulations require that all drivers of vehicles carrying dangerous goods (above certain limits):

- Must attend and pass an approved basic training course.
- Specialised additional training for tanker drivers and for specific substances (explosives and radioactives).
- Refresher training every five years.

Some subjects covered in basic training:

- General requirements of DG Regs/ADR.
- Types of hazards.
- Preventive and safety measures.
- Emergency action.
- Marking/labelling/placarding.
- Loading/unloading precautions.

MORE...

Further information on carriage of dangerous goods can be obtained from the HSE at:

www.hse.gov.uk/cdg

ADR

Article 2 of the current version of the ADR Agreement states that, apart from some excessively dangerous goods, other dangerous goods may be carried internationally in road vehicles subject to compliance with:

- the conditions laid down in Annex A for the goods in question, in particular in regard to their packaging and labelling; and
- the conditions laid down in Annex B, in particular in regard to the construction, equipment and operation of the vehicle carrying the goods in question.

HINTS AND TIPS

You should note that Annexes A and B have been regularly amended and updated since the entry into force of ADR, so it is important to use a current copy.

Carriage of Dangerous Goods and Use of Transportable Pressure Equipment Regulations 2009

These Regulations (and the **Carriage of Dangerous Goods and Use of Transportable Pressure Equipment (Amendment) Regulations 2011**) control the carriage of dangerous goods by road and rail in Great Britain and also, in so far as it relates to safety advisers, regulate the carriage of dangerous goods by inland waterway. The UK is required to apply the provisions of ADR and RID to national transport within its territory because EU Directives require ADR and RID to be applied (although the UK, like other member states of the European Community, is permitted to modify in certain ways how ADR and RID are applied within its territory).

MORE...

The current copies of the UN Recommendations (Orange Book), the *Manual of Tests and Criteria* and ADR can be viewed and downloaded free from the Legal Instruments and Recommendations area of the UN's Dangerous Goods web pages:

https://unece.org/transport/dangerous-goods

The key areas to briefly note in the Regulations are:

- **Requirements of ADR and RID** which address issues such as:
  - Training requirements.
  - Compliance with safety obligations.
  - Special requirements relating to the carriage of class 7 (radioactive) goods.
  - Appointment of Dangerous Goods Safety Advisers.
  - Reporting accidents or incidents.
  - Security provisions.
  - Requirements relating to the construction and testing of packaging, receptacles and containers.
  - Carriage, loading, unloading and handling.
  - Requirements for vehicle crews, equipment, operation and documentation.
  - Construction and approval of vehicles.
- **Requirements in addition to ADR and RID** which include:
  - Additional security requirements for carriage of class 1 (explosive) goods by road.
  - Keeping of consignment information.
  - Alternative placarding, marks and plate markings for carriage within Great Britain.
- **Transportable pressure equipment** which specifies:
  - Design and manufacturing standards.
  - Requirements for periodic inspection.

The **Carriage Regulations** reinforce the ADR requirement (in Chapter 8.2) for drivers of vehicles carrying dangerous goods to attend a vocational course of instruction and sit an externally assessed examination - for the classes of goods carried (as usual there are some exemptions for small quantities). This gives them a certificate of competency, which is required to be updated at specific intervals.

The **Transport of Dangerous Goods (Safety Advisers) Regulations 1999** (via Chapter 1.8.3.1 of ADR) require the appointment of a qualified safety adviser or advisers (usually referred to as a 'Dangerous Goods Safety Adviser' or 'DGSA'). The DGSA must hold a vocational training certificate which covers the modes of transport used and the classes of dangerous goods transported by the employer.

The main duties of the safety adviser are:

- Monitoring legal compliance requirements on the transport of dangerous goods.
- Ensuring that an annual report is prepared on the activities of the employer which concern the transport of dangerous goods.

- Providing advice to the employer on the transport of dangerous goods.
- Monitoring the employer's arrangements for:
  - Identification of dangerous goods.
  - Requirements for transport vehicle purchase.
  - Checking transport equipment.
  - Training employees.
  - Implementing emergency procedures.
  - Investigation and preparation of reports on serious accidents.
  - Implementing remedial action following an accident.
  - Ensuring compliance with regulations when choosing and using sub-contractors or third parties.
  - Verifying that employees involved in carriage, loading, etc. of dangerous goods have detailed procedures and instructions.
  - Introducing measures to increase risk awareness in relation to dangerous goods activities.
  - Implementing verification procedures to ensure necessary documents and emergency equipment are present on board vehicles.
  - Implementing verification procedures to ensure compliance with requirements concerning loading/unloading.
  - Ensuring existence of a security plan.

The safety adviser must also ensure that a report is prepared on any accident affecting health and safety that occurs during the transport of dangerous goods.

STUDY QUESTIONS

3. What features should be considered when assessing the suitability of a store for flammable liquids?
4. Describe the main means of spillage containment.
5. What measures should be taken to avoid overfilling with hazardous liquid?
6. In accordance with the **Dangerous Substances and Explosive Atmospheres Regulations**, which zone would spray-paint booths fall under?
7. When selecting a container for filling, what important factors should be checked?
8. Why should dispensing of flammable liquids be done in a well-ventilated area?
9. How can unloading of a substance from a tanker into the wrong storage tank be prevented?

(Suggested Answers are at the end.)

Hazardous Environments

IN THIS SECTION...

- Where electrical equipment may be exposed to hazardous environments, it should be constructed and protected to prevent danger arising from the exposure. This should include:
  - Resistance to mechanical damage.
  - Protection against solids, dusts, liquids and gases.
  - Protection against other hazards, e.g. wet environments.
- Where flammable atmospheres are present (or likely to occur), particular precautions are necessary to prevent electrically-caused ignition of the atmosphere. Hazardous area classification identifies those areas where flammable atmospheres can be found and provides an estimate of how often they may be found there.
- Zones where flammable atmospheres could occur are defined on the basis of the likelihood and frequency of the presence of a flammable atmosphere:
  - Zones 0, 1 and 2 for gases, vapours and mists.
  - Zones 20, 21 and 22 for dusts.
- Electrical equipment intended for use in such a zone must be of a suitable category. Three categories exist in international standards; numbered 1, 2 and 3.
- Within these categories a number of different methods are used to ensure safety, these methods include intrinsic safety, flameproof enclosures and pressurised apparatus.
- Work involving electrical systems in hazardous environments, such as those which are flammable, corrosive, explosive, or involving Zones 0, 1 and 2, should be carried out in accordance with a permit-to-work system.

Introduction to Hazardous Environments

Where electrical equipment may be exposed to adverse or hazardous environments, the equipment should be constructed and protected to prevent danger arising from the exposure. The protection required will vary depending on the type of hazard and the degree of risk. The foreseeable risks associated with electrical equipment in hazardous environments are:

Mobile phones can act as an energy source and may ignite flammable vapours

- Degradation of the electrical equipment by mechanical damage, chemical attack, corrosion or simply water ingress so that the risk of electric shock exists. For example, an unprotected electrical cable laid across a vehicle traffic route might be physically crushed so that the insulation is degraded to the point where touching the cable would result in electric shock.
- The equipment acting as an energy source, causing ignition of an explosive atmosphere such as a flammable gas, mist or vapour or an explosible dust. For example, a normal mobile phone carried into an area where extremely flammable liquid is being handled and decanted might ignite the flammable vapours, causing an explosion.

Regulation 6 of the **Electricity at Work Regulations 1989 (EAWR)** is concerned with adverse or hazardous environments.

Principles of Protection

Resistance to Mechanical Damage

Electrical equipment, installations and systems, and in particular cable runs, must be adequately protected from the foreseeable mechanical damage that they might be subject to. This includes abrasion, impact, stress, wear and tear, vibration and hydraulic and pneumatic pressure.

Protection from mechanical damage can be provided to a degree by placing the equipment/system in a safe location. For example, running power cables overhead protects them from being run over by, or impacted by, vehicular or pedestrian traffic; placing an electrical distribution board for a warehouse in an adjacent service room protects the installation from impact by forklift trucks operating in the warehouse.

Alternatively, it may be possible to protect electrical installations using a physical barrier, such as posts or guardrails that prevent impacts. For example, a substation inside a palisade fence, adjacent to a lorry manoeuvring yard, might be further protected from vehicle impact if a heavy duty traffic barrier between the substation and the yard was installed.

In many cases, mechanical protection is offered by the:

- design and manufacture of the electrical equipment or cable itself; or
- running the cable inside an appropriately specified conduit or trunking.

Cables can incorporate multiple layers of insulation surrounding their inner cores, and these outer layers can incorporate wire that acts as a protective cage. In effect, armouring the cable against mechanical damage. This is often used as a means of protection for both low- and high-voltage cables; however, it does not provide total protection from mechanical damage. The striking of buried, high-voltage cables with excavators is a major cause of fire and explosion on construction sites.

One of the most significant means of protecting cables from mechanical damage is to place the cable inside a protective conduit or trunking. Typical materials used for these include:

- Plastic - for light-duty applications such as domestic systems.
- PTFE - for medium-duty installations and those where better temperature and chemical resistance is required.
- Metal - for a high level of mechanical protection.
- Concrete ducts - for outdoor or buried systems.

The level of mechanical protection required is subject to legislation and standards; mechanical protection of electrical systems is:

- required by the **Electricity at Work Regulations 1989**;
- with the relevant standards set out in the **Institute of Engineering and Technology (IET) Wiring Regulations**; and
- supported by various internationally recognised technical standards such as EN 62262 (degrees of protection provided by enclosures for electrical equipment against external mechanical impacts).

Protection Against Solid Objects, Dusts, Liquids and Gases

Electrical equipment, installations and systems need to be suitable for the environment in which they are placed or used. Two significant environmental factors that might make an electrical system unsafe are the possibility of solid objects penetrating the enclosure of the electric equipment, especially small solid materials such as grit, dirt or dust, and the possibility of liquid ingress, especially water ingress into the enclosure. Solid and liquid ingress might cause:

- direct mechanical damage to the electrical system (e.g. a mechanical switch might be prevented from opening or closing due to dust ingress);
- corrosion damage (e.g. contacts might become so badly corroded by water ingress that they become non-conducting); or
- the installation to become unsafe by providing a conductive material that allows stray current to flow (including the very real possibility of electric shock and fire).

The electrical equipment, therefore, must be carefully selected so as to be suitable for the environment of use. This is done by ensuring that equipment is manufactured to a relevant specification. A good example of this is the use of the **Index of Protection (IP) rating system**.

The IP rating system is a standard (60529) published by the International Electrotechnical Commission (IEC). The rating system uses two numbers to give an indication of the degree of protection that an electrical enclosure has against both solid object/dust ingress and water ingress. These numbers follow the prefix IP - thus IP11, IP35, IP56, etc.

The first number gives an indication of solid/dust protection from 0 (no protection) to 6 (dust tight; no ingress of dust and complete protection against contact).

The second number gives an indication of water protection from 0 (no protection) to 8 (the equipment can be immersed in water at depths of 1m or more).

Thus, the rating IP66 means that the electrical equipment is dust tight and can withstand exposure to powerful water jets from any direction.

The IP rating is affixed by the manufacturer of the electrical equipment and represents their assurance that the equipment can withstand exposure to dust and water in circumstances represented by the standard test methods set out in the IET standard. An employer can use these ratings to make an appropriate selection for the electrical equipment they intend to install in a particular environment.

Wet Environments

In certain workplaces, e.g. those carrying out electroplating activities, there may be significant use of conductive and corrosive fluids which can generate humid and corrosive atmospheres. In this type of environment, there will be the risk of corrosion and subsequent damage of any uninsulated or unprotected parts of the electrical system, and also the possibility of ingress of fluids or water vapour into electrical components.

In such an environment, it is essential that all electrical equipment is designed and constructed to withstand this type of hazardous environment.

Protection Against Other Hazards

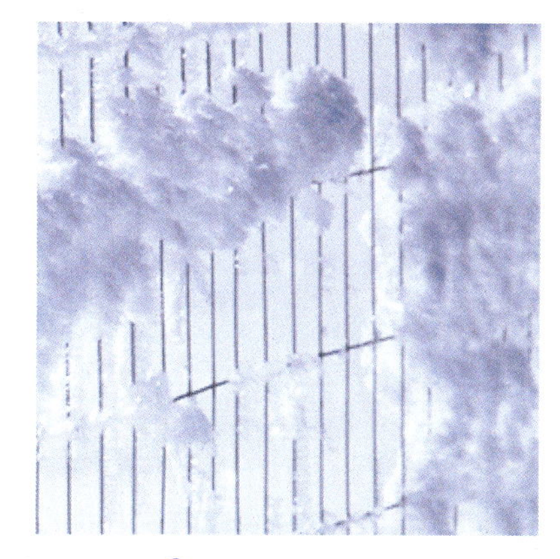

Snow can create a hazardous environment

Adverse weather conditions, including extreme temperature, rain, snow, ice and wind, can create a hazardous environment. Additional protection may include lightning protection and installing weatherproof equipment. It should be ensured that all equipment exposed to external conditions is sufficiently robust to cope with the hazards likely to be encountered.

Natural hazards also include solar radiation, animals and plants. Sufficient protection may be achieved by simply increasing the strength of equipment, such as insulation that can withstand the gnawing of rodents.

Classification of Hazardous Areas and Zoning

Hazardous area classification identifies those areas where flammable atmospheres may be found and provides an estimate of how often they may be found there. It is applied to areas where solvent vapours, gases or mists may exist and also those where flammable dust clouds can occur.

The principle of hazardous area classification has featured in various national standards for a long time. In all EU countries, it is mandatory as the **ATEX Directives** (Atmosphères Explosibles) are incorporated into the national laws of all member states. Regulation 7(2) of **DSEAR** requires employers to classify places at the workplace where an explosive atmosphere may occur into hazardous or non-hazardous places in accordance with Schedule 2 of **DSEAR**. The zones where flammable atmospheres could occur are defined on the basis of the likelihood and frequency of the presence of a flammable atmosphere:

- Zone 0, 1 and 2 for gases, vapours and mists.
- Zone 20, 21 and 22 for dusts.

Although the zone designations are different for gases and dusts, the actual definitions are essentially identical, as shown in the next section.

Hazardous Area Zoning

Here's a reminder of the zone classifications set out in **DSEAR**:

Flammable Vapours

- **Zone 0**

  A place in which an explosive atmosphere consisting of a mixture with air of dangerous substances in the form of gas, vapour or mist is **present continuously or for long periods or frequently**.

- **Zone 1**

  A place in which an explosive atmosphere consisting of a mixture with air of dangerous substances in the form of gas, vapour or mist is **likely to occur in normal operation occasionally**.

- **Zone 2**

  A place in which an explosive atmosphere consisting of a mixture with air of dangerous substances in the form of gas, vapour or mist is **not likely to occur in normal operation but**, **if it does occur**, **will persist for a short period only**.

Combustible Dusts

- **Zone 20**

  A place in which an explosive atmosphere in the form of a cloud of combustible dust in air is **present continuously, or for long periods or frequently**.

- **Zone 21**

  A place in which an explosive atmosphere in the form of a cloud of combustible dust in air is **likely to occur in normal operation occasionally**.

- **Zone 22**

  A place in which an explosive atmosphere in the form of a cloud of combustible dust in air is **not likely to occur in normal operation but, if it does occur, will persist for a short period only**.

Determining which zone is suitable for different circumstances, and where the boundaries between zones should be drawn, is a complex process which should involve expertise from a number of fields such as safety, production and chemical engineering.

Once the zone boundaries are established, equipment can be selected so that it is suitable for the intended area.

Use of Permits to Work

Permit-to-Work Procedures

A Permit-To-Work (PTW) system is a formal recorded process used to control work which is identified as potentially hazardous. It is also a means of communication between site/installation management, plant supervisors and operators and those who carry out the hazardous work.

Permits to work are needed for work of a potentially hazardous nature

Whenever maintenance or other temporary work of a potentially hazardous nature is to be carried out, some sort of permit-to-work system is essential and work on electrical equipment or electrical systems in hazardous environments fall into this category.

PTW procedures are used to ensure that non-routine, usually hazardous work is assessed, planned, authorised and carried out in such a way as to ensure the health and safety of the workers involved, and others who may be affected. It ensures that proper consideration is given to the risks and that they are dealt with before the task starts and throughout the duration of the ongoing work. Equally important is controlling the completion of the work.

In most cases, a PTW system should be used to control maintenance operations in hazardous environments. PTWs are formal management documents and should only be issued by those with clearly assigned authority to do so. The requirements stated in them must be complied with before the permit is issued and the work covered by it is undertaken.

Individual PTWs need to relate to clearly defined individual pieces of work.

PTWs should normally include:

- the location and nature of the work intended;
- identification of the hazards, including the residual hazards and those introduced by the work itself;
- the precautions necessary, for example, isolations or gas testing;
- the personal protective equipment required;
- the proposed time and duration of the work;
- the limits of time for which the permit is valid; and
- the person in direct control of the work.

Through this process the potential effects of the hazardous environment on the electrical equipment or system in question will be identified and the necessary precautions required will be incorporated into the permit system.

Selection of Electrical Equipment for Use in Flammable Atmospheres

In situations where flammable atmospheres are present or likely to occur occasionally, particular precautions are necessary to prevent electrically-caused ignition of the atmosphere. Hot surfaces, arcs and sparks associated with electrical equipment can ignite such atmospheres, causing fires and explosions. To prevent such ignition, fire and explosion:

- Areas where such flammable atmospheres might exist must be recognised.
- Electrical equipment for use in such areas must be selected to ensure that it is suitable and will not cause fire or explosion.

Flammable atmospheres are created by the presence of flammable gases, vapours and dusts and for an atmosphere to be capable of ignition, the concentration of the flammable substance in air must be at a certain level - within the Flammable or Explosive Range which lies between the upper and lower explosive limits (UEL and LEL) of the substance. Mixtures outside this range, i.e. rich and lean mixtures, will not ignite.

Pressurisation and Purging

Purge and pressurisation is a two-step process carried out prior to energising the electrical equipment inside an enclosure. The goal is to ensure that once an enclosure is purged and pressurised with a protective gas supply, the enclosure be energised or powered up.

The protective gas supply needs to be free of any hazardous or explosive traces and have the capacity to sustain the purge and pressurisation process. In some special applications, an inert gas is used as the protective gas supply (argon, nitrogen, or a mixture of inert gases) instead of a standard atmospheric air mixture.

Pressurisation is the process of creating a higher internal pressure that is provided by a protective gas supply, preventing any hazardous gas or dust from entering the enclosure. Any penetrations or leak areas will have protective gas exiting the enclosure rather than hazardous gas or dust migrating into it. This segregates any external explosive or hazardous material from the energised internal equipment.

Purge is the process used to remove any potentially hazardous gas from the interior of the enclosure prior to pressurisation. The purge cycle performs 'air-exchanges' that displace any explosive (hazardous) gas with inert, protective gas instead.

Once completed, all potentially explosive gas has been removed from the enclosure's interior. This can be done either manually or automatically.

Energisation of the enclosure occurs only after the purge and pressurisation process has eliminated the potential for internal explosive gas inside the enclosure.

The completion of this process makes the interior of the enclosure safe for the appropriate devices to be energised.

Categories of Electrical Equipment

Electrical equipment for use in zoned atmospheres is classified by the manufacturer into one of three categories: category 1, 2 or 3:

- **Category 1** equipment is suitable for use in a Zone 0, 1 or 2 area or a Zone 20, 21 or 22 area.
- **Category 2** equipment is suitable for use in a Zone 1 or 2 area or a Zone 21 or 22 area.
- **Category 3** equipment is suitable for use in a Zone 2 area or a Zone 22 area.

Thus, Category 1 equipment gives the highest level of protection and category 3 the lowest. These standards are set out in the **Equipment and Protective Systems Intended for Use in Potentially Explosive Atmospheres Regulations 2016**. Schedule 3 of **DSEAR** states that equipment and protective systems must be selected on the basis of these regulations.

The methods used by the manufacturer of the electrical equipment to make it safe to use in the zoned environment vary but fall into a number of different techniques, each of which is indicated by a 'type of protection' code on the electrical equipment itself:

Type of protection	Basic principle
Flameproof enclosure **(d)**	Parts which can ignite a potentially explosive atmosphere are surrounded by an enclosure which can withstand the pressure of an explosive mixture exploding inside of it and prevents the propagation of the explosion to the atmosphere surrounding the enclosure. It can be used in Zones 1 or 2.
Increased safety **(e)**	Additional measures are taken to increase the level of safety, thus preventing the possibility of unacceptably high temperatures and the creation of sparks or electric arcs within the enclosure or on exposed parts of electrical apparatus, where such ignition sources would not occur under normal operation.
Pressurised apparatus **(p)**	The formation of a potentially explosive atmosphere inside a casing is prevented by maintaining a positive internal pressure of inert gas in relation to the surrounding atmosphere and, where necessary, by supplying the inside of the casing with a constant flow of inert gas which acts to dilute any combustible mixtures.*
Intrinsic safety **(i)**	Apparatus used in a potentially explosive area contain intrinsically safe electric circuits only. An electric circuit is intrinsically safe if no sparks or thermal effects are produced under specified test conditions (which include normal operation and specific fault conditions) which might result in the ignition of a specified potentially explosive atmosphere.
Oil immersion **(o)**	Electrical apparatus (or parts of it) are immersed in a protective fluid (such as oil), such that a potentially explosive atmosphere existing over the surface or outside of the apparatus cannot be ignited.
Powder filling **(q)**	Filling the casing of an electrical apparatus with a fine granular packing material, which makes it impossible for an electric arc created within the casing under certain operating conditions to ignite a potentially explosive atmosphere outside of it. Ignition must not result from either flames or raised temperature on the surface of the casing.

Type of protection	Basic principle
Encapsulation (m)	Parts which may ignite a potentially explosive atmosphere are embedded in sealing compound, such that the potentially explosive atmosphere cannot be ignited.
Type of protection (N)	Electrical apparatus is not capable of igniting a potentially explosive atmosphere (under normal operation and under defined abnormal operating conditions).

'Intrinsically safe' means that the energy level of the equipment is insufficient to produce an incendiary spark. Two categories of intrinsically safe equipment exist:

- 'ia' which is more stringent as it allows for two simultaneous faults; and
- 'ib' which allows for only one.

Only 'ia' equipment can be used (exceptionally) in Zone 0 if sparking contacts are not part of the equipment. Examples of type 'i' are instrumentation and low-energy equipment.

The type of protection appears on the electrical equipment alongside an EEX precursor, an Ex symbol and the category of the equipment.

Thus, Category 2 equipment of Type ib is suitable for use in a Zone 1 or 2 area (or 21 or 22) and is intrinsically safe.

STUDY QUESTIONS

10. How should electrical equipment, which is liable to be damaged by corrosion from moisture, be constructed?
11. Describe flameproof equipment and identify the zones for which it is suitable.
12. Describe what is meant by 'purging'.
13. Describe a hazardous area classified as 'Zone 2'.

(Suggested Answers are at the end.)

Emergency Planning

IN THIS SECTION...

- Organisations need procedures in place to cope with emergencies. Typical emergencies might include:
  - Localised chemical spillage or gas/vapour release.
  - Fire evacuation.
  - First-aid treatment.
  - Bomb threat.
  - Major incident (such as a large chemical vapour release).
- In the event of an emergency, access to first-aid personnel and facilities, means for fighting a fire to ensure that it is extinguished before any serious damage is done, and spill containment to prevent or reduce damage or injury, can mean the difference between life and death.
- An emergency plan is a formal, written document designed to assist management with the control of specific hazards or incidents, to minimise disruption to normal work activities and reduce the impact on the organisation, including post-incident recovery.
- The **Control of Major Accident Hazards Regulations 2015 (COMAH)** apply to special risk premises and require that sites to which the Regulations apply have emergency plans. The Regulations make a distinction between:
  - 'upper-tier' sites, which hold large quantities of dangerous substances; and
  - 'lower-tier' sites, where the quantities are significantly less.
- Upper-tier operators need to:
  - Prepare and test an **on-site (internal) emergency plan** - this lays down the response to the emergency by those on the site.
  - Provide local authorities with information to enable them to prepare off-site (external) emergency plans - this lays down the co-ordinated response of external agencies to a site emergency, which may have off-site effects.
  - Review, and where necessary revise, their emergency plans to take account of any changes in the site, or within the emergency services concerned, new technical knowledge, or knowledge concerning the response to major incidents.

Need for Emergency Preparedness Within an Organisation

Emergencies can take many forms and can occur on different scales. Organisations need procedures in place to cope with emergencies. Typical emergency procedures that a company might need to consider include:

- Localised chemical spillage or gas/vapour release.
- Fire evacuation.
- First-aid treatment.
- Bomb threat.
- Major incident (such as a large chemical vapour release).

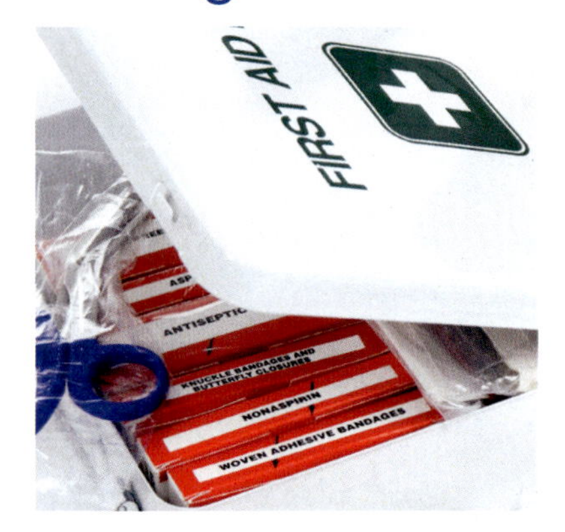

Emergency situations can occur at any time. How an organisation deals with that emergency, however, can dictate whether the emergency turns into a disaster or not. Historically, there have been a number of disasters that could easily have been prevented or where the consequence could have been much less if appropriate safeguards or emergency planning had been in place.

For these reasons it is imperative that all organisations prepare themselves for emergency situations. Action should be considered in terms of:

- Prevention of loss.
- Early warning of loss.
- Minimising the consequences of loss.

The requirement for emergency planning is included in many pieces of health and safety legislation (either express or implied), including:

- **Management of Health and Safety at Work Regulations 1999 (MHSWR)**
  - Preventive and protective measures (Reg. 5).
  - Procedures for serious and imminent danger (Reg. 8).
  - Contacting external services in case of emergency (Reg. 9).
- **Control of Substances Hazardous to Health Regulations 2002 (COSHH)**

  Arrangements to deal with accidents, incidents and emergencies (Reg. 13).
- **Dangerous Substances and Explosive Atmospheres Regulations 2002 (DSEAR)**

  Arrangements to deal with accidents, incidents and emergencies (Reg. 8).
- **Control of Major Accident Hazards Regulations 2015 (COMAH)**

  For premises with a special risk (which we shall look at later).

Consequence Minimisation via Emergency Procedures

First Aid/Medical Provision

The **Health and Safety (First-Aid) Regulations 1981** require employers to provide adequate and appropriate equipment, facilities and personnel to ensure their employees receive immediate attention if they are injured or taken ill at work. It has to be noted that these Regulations apply to all workplaces **including those with less than five employees and to the self-employed**.

Employers are required to carry out an assessment of first-aid needs and this will involve consideration of all the workplace hazards and the risks arising, the size of the organisation and any other relevant factors, to determine what first-aid equipment, facilities and personnel should be provided.

In the event of an emergency, access to good first-aid personnel and facilities is essential and while most organisations will have a number of qualified first aiders, it is desirable for as many employees as possible to have some first-aid training in, for example, artificial respiration, control of bleeding and any other relevant action for hazards in their own particular workplace.

It is important that qualified first aiders are identified on a notice which is displayed in a prominent place in the workplace. Contact numbers and times should also be displayed and these should be kept up to date. Apart from their vocational training, first aiders should also be trained in any relevant or specific chemicals, plant or equipment used in the workplace and be able to deal with injuries arising from them such as chemical burns, trauma or crush injuries.

First-aid equipment suitable to the workplace should be available in an easily accessible location. Usually first-aid boxes are adequate and these should be strategically situated around the workplace. Other first-aid items, such as eye washes, may be necessary where there is a particular hazard from dust or airborne debris.

More complex establishments might have a large cohort of first-aid trained personnel and perhaps a dedicated first-aid room with facilities for treating quite serious injuries while an office environment will only require a small first-aid kit kept in the managers office.

It is vital that all first-aid equipment is checked regularly and replenished as necessary.

The contents of a first-aid kit should be based on the company's first-aid needs assessment and should comply with BS 8599, although this is not a legal requirement.

MORE...

Further information on first-aid requirements and advice can be found here:

www.hse.gov.uk/pubns/indg347.htm

Fire Evacuation

Studies of means of escape in Learning Outcome 11.4 have established that to **protect the occupants**, a building must incorporate features which will protect them **while they escape from a fire**. Likewise, the building must not incorporate features which would endanger the occupants or prevent their escape in the event of a fire. The main factors to consider in provision and maintenance of means of escape are the:

- Nature of the occupants, e.g. mobility.
- Number of people attempting to escape.
- Distance they may have to travel to reach a place of safety.
- Size and extent of the 'place of safety'.

The total of these considerations results in an **adequate means of escape**, by which, under the worst conditions, the exposure of the occupants to danger will be minimal.

To assist employees, visitors and contractors to escape in the event of a fire, comprehensive Fire Evacuation Plans should be in place for even the simplest of workplaces.

As a minimum, emergency procedures should:

- Describe how to:
  - Raise the alarm and call the fire and rescue service.
  - Tackle a fire or control spills and leaks.
  - Evacuate the site.
- Include arrangements for evacuation of areas in the event of fire or toxic gas emission and specify designated safe areas, assembly points and toxic gas shelters.
- Identify responsible personnel whose duties during area evacuation include:
  - Responsibility for a specific area.
  - Collecting ID badges from plant racks.
  - Ensuring roll calls are undertaken to identify missing persons.
  - Communication of missing persons to central emergency services.

Regular fire evacuation drills should be carried under realistic conditions with a 'wash up' meeting of all parties concerned at the end to discuss lessons learned and implement any changes to the plans that might be required.

Spill Containment

There are a number of ways in which a spill can be contained so as to prevent or reduce damage or injury. Most of these relate to the design of the vessel or pipe or surrounding area.

Pressure relief and emergency venting provides protection against over-pressurisation of plant, and may be the last line of defence against failure and uncontrolled loss of containment. Adequate isolation valves should also be provided so that in the event of a sudden loss the effects can be mitigated. Where personnel would be exposed to danger when operating the valves manually, the shut-off valves should be remotely operated wherever reasonably practicable.

Suitable bunding around storage vessels will capture some loss safely and act as an indication that the system requires review so as to prevent further loss.

Response to an emergency spill

In addition to the above, good housekeeping in terms of regular checks of the vessels, pipework and any safety features and preventive maintenance should ensure that unplanned spills are prevented.

Facilities that are subject to the **COMAH Regulations** should have procedures in place for dealing with emergency situations involving loss of containment by:

- containing and controlling incidents so as to minimise the effects and to limit danger to persons, the environment and property;
- implementing the measures necessary to protect persons and the environment;
- description of the actions which should be taken to control the conditions at events and to limit their consequences, including a description of the safety equipment and resources available;
- arrangements for training staff in the duties they will be expected to perform;
- arrangements for informing local authorities and emergency services; and
- arrangements for providing assistance with off-site mitigatory action.

The emergency plan should be simple and straightforward, flexible and achieve necessary compliance with legislative requirements. Furthermore, separate on-site and off-site emergency plans should be prepared (see later in this Learning Outcome).

The emergency spill control procedure should include the following key sections:

- Spills involving hazardous materials should first be contained to prevent spread of the material to other areas. This may involve the use of temporary diking, sand bags, dry sand, earth or proprietary booms/absorbent pads.
- Wherever possible, the material should be rendered safe by treating with appropriate chemicals or diluted to a safe condition.
- Hazardous materials in a fine, dusty form should not be cleared up by dry brushing. Vacuum cleaners should be used in preference, and for toxic materials one conforming to type H should be used.
- Treated material should be absorbed onto inert carrier material to allow the material to be cleared up and removed to a safe place for disposal or further treatment as appropriate.
- Waste should not be allowed to accumulate. A regular and frequent waste removal procedure should be adopted.

MORE...

Further guidance can be found at:

www.hse.gov.uk/pubns/books/l111.htm

Development and Maintenance of Emergency Plans to Meet Regulatory Requirements

The **COMAH Regulations** apply to special risk premises. These Regulations require that sites to which they apply have emergency plans. The Regulations make a distinction between 'upper-tier' sites, which hold large quantities of dangerous substances (e.g. LPG, petroleum spirit, explosives, oxygen and specified toxic substances), and 'lower-tier' sites where the quantities are significantly less. The 'competent authority' (i.e. the body that enforces regulations) is jointly the HSE and the Environment Agency (Scottish Environment Protection Agency in Scotland, Natural Resources Wales in Wales).

Development and Maintenance of Emergency Plans

Following the fire and explosion at the Buncefield oil storage depot in 2005, the Buncefield Major Incident Investigation Board made a recommendation that the **COMAH** Competent Authority (CA) improve its guidance for producing on- and off-site emergency plans.

The purpose of the **COMAH Regulations** is to ensure that the consequences of a major accident at a **COMAH** site are minimised through the provision of effective on-site emergency planning and response arrangements and where necessary, dovetailing with the off-site emergency plans prepared by the local authorities under **COMAH** or Civil Contingencies legislation.

Under **COMAH**, operators should prepare an on-site emergency plan for dealing with the on-site consequences of possible major accidents and providing assistance with off-site mitigation measures. An off-site emergency plan addresses major accidents that have off-site consequences to persons and the environment.

Lower-tier operators are required by **COMAH** Regulation 4 to take all measures necessary to prevent major accidents and limit their consequences to persons and the environment. Regulation 5 requires operators to prepare a Major Accident Prevention Policy (MAPP) (see later), including planning for emergencies and adoption of procedures to:

- identify foreseeable emergencies,
- prepare, test and review emergency plans, and
- provide training for all persons working in the establishment.

The operator needs to consider reasonably foreseeable, low-probability, high-consequence events. The level of planning for these emergencies should be proportionate to the probability of their occurring.

The procedures required by the safety management system must ensure that an adequate emergency plan is developed, adopted and implemented. The procedures should also cover the necessary arrangements for communicating the plans to those likely to be affected by an emergency; in particular, information sharing with any off-site emergency responders.

An emergency plan is a formal, written document designed to assist management with the control of specific hazards or incidents, so that minimum disruption to normal work activities will occur and the good name of the company will not be damaged.

When the range of major disruptive circumstances which could arise have been identified, individual emergency plans will then cover the following main points with regard to each of the identified hazards:

- Event.
- Location.
- Potential for harm.
- Existing instructions for dealing with the problem.
- Immediate actions to be taken.
- Control of the event.
- Assessment of the event.
- Response.
- Damage limitation action.
- Recovery plan.

TOPIC FOCUS

The issues that need to be considered in the **development of an emergency plan to minimise the consequences of a major incident in a chemical plant using toxic and flammable substances** include:

- Causes and extent of damage from a major incident.
- Quantities of materials stored.
- Call-out arrangements.
- Resources to deal with the incident including first-aid and rescue equipment.
- Raising the alarm.
- On- and off-site evacuation and shelter arrangements.
- Staff training (induction and refresher).
- Notification of, and coordination/consultation with, emergency services.
- Search and rescue arrangements.
- Communication issues.
- Business continuity arrangements.
- Control of spillages and pollution and clean-up and decontamination.
- Establishing control centres and making available information, plans and inventories.

Duties of Site Operators

Lower-Tier Sites

Operators must notify the competent authority before operations with dangerous substances begin or within three months of an establishment becoming subject to the Regulations. They must also take all measures necessary to prevent major accidents and limit their consequences to people and the environment. Prevention should be based on the principle of reducing risk to a level that is '**As Low As is Reasonably Practicable**' (**ALARP**) for human risks, and using the '**Best Available Technology Not Entailing Excessive Cost**' (**BATNEEC**) for environmental risks. The aim is always to avoid the hazard altogether where possible.

The key documentation that needs to be prepared is the **Major Accident Prevention Policy** (**MAPP**). This demonstrates how the safety management systems will prevent major accidents, and consequently covers areas such as:

- Organisation and personnel.
- Identification and evaluation of major hazards.
- Operational control.
- Planning for emergencies.
- Monitoring, audit and review.

This may be quite short, being little more than a policy together with supporting information to show that a safety management system has been established.

MAPP documents must be provided to the competent authority as soon as possible and in any case within three months of the establishment becoming subject to the Regulations.

Upper-Tier Sites

This is similar to above; however, operators must also prepare a comprehensive **safety report** (which includes a MAPP). The safety report provides information to demonstrate to the competent authority that all measures for the prevention and mitigation of major accidents have been taken.

The safety report must include:

- Information on the management system and on the organisation of the establishment with a view to major accident prevention (i.e. a MAPP).
- Layout of establishment (maps of site, geography, identification of major hazard installations or activities).
- Description of installation (main activities, major accident risks, processes, operating methods, dangerous substances, including types and amounts).
- Detailed description of major accident scenarios, analysis of probability (or circumstances), extent and severity of consequences, description of safety precautions built into installation (operating parameters, venting, design, etc.). It is worth revising the section on the use of Quantified Risk Assessment in Unit 1 Learning Outcome 4).
- Information on protective measures and intervention to limit consequences of any major accident that might occur:
  - Emergency procedures - equipment installed (or available).
  - People and resources that can be mobilised and how they will be used to carry out the on-site emergency plan.
  - Summary of information needed to draw up the emergency plan for the site.

The safety report must be provided "a reasonable period of time" prior to construction of the establishment and in any case within one year of the establishment becoming subject to the Regulations. It needs to be kept up to date and reviewed every five years or after significant changes.

In addition, upper-tier operators also need to:

- Prepare and test an **on-site (internal) emergency plan** - this lays down the response to the emergency by those on the site.
- Provide local authorities with information to enable them to prepare **off-site (external) emergency plans** - this lays down the co-ordinated response of external agencies to a site emergency, which may have off-site effects.

Members of the public, schools, hospitals, etc. who could be affected by an accident at a **COMAH** establishment must also be provided with information, and updates "supplied regularly".

On-site emergency plans must be prepared before the site starts to operate (for one which hasn't yet started to operate) or within one year of the establishment becoming subject to the Regulations (for establishments which are already operating and have fallen within the scope of the Regulations). They must be tested and reviewed at least every three years and revised as necessary.

Off-site emergency plans must also be tested and reviewed and changes made as necessary. Local authorities must consult with "appropriate" members of the public during any such review.

Control room of a major hazard site

We can see from the outline above that the aim is to develop an effective management system, with adequate controls to **prevent** a major accident, then ensure that effective emergency plans are in place to **mitigate** the consequences if an accident does occur. The process includes:

- Identifying potential accident sequences.
- Introducing precautions to reduce the likelihood.
- Assessing the impact off-site if the accident does occur.
- Identifying how the impact can be minimised.
- Providing resources and systems to control the event.

Preparation of On-Site and Off-Site Emergency Plans Including Monitoring and Maintenance

The two emergency plans should be complementary. The on-site plan should include details of the arrangements in place to assist with an emergency off-site. Similarly, the off-site plan should include details of the arrangements for providing assistance for the on-site emergency.

The emergency plan should cover what response is required during each phase of the emergency, both immediately and in the longer term. During the first few hours after the accident, the 'critical' phase of an accident response, key decisions must be made quickly and under considerable pressure. A detailed understanding of the likely sequence of events and appropriate actions will help anyone who may be expected to play a part in the response.

In developing the safety report required by **COMAH**, the operators of upper-tier establishments need to describe the:

- Equipment that is installed or available in the plant to limit the consequences of a major accident.
- Number of people and resources that can be mobilised and how they will be used to carry out the on-site emergency plan.

Some of the information needed for the emergency plan is therefore included in the operator's safety report (for upper-tier **COMAH** sites).

On-Site

Part 1 of Schedule 4 to **COMAH** identifies the information that should be included in an **internal emergency plan (on-site)**:

- The names or positions of persons authorised to set emergency procedures in motion, and the person in charge of and co-ordinating the on-site mitigatory action.
- The name or position of the person with responsibility for liaison with the local authority responsible for preparing the off-site emergency plan.
- For foreseeable conditions or events which could be significant in bringing about a major accident, a description of the action which should be taken to control the conditions or events and to limit their consequences, including a description of the safety equipment and the resources available.
- The arrangements for limiting the risks to persons on site, including how warnings are to be given and the actions persons are expected to take on receipt of a warning.
- Arrangements for providing early warning of the incident to the local authority responsible for setting the off-site emergency plan in motion, the type of information which should be contained in an initial warning, and the arrangements for the provision of more detailed information as it becomes available.
- Arrangements for training staff in the duties they will be expected to perform, and where necessary co-ordinating this with the emergency services.
- Arrangements for providing assistance with off-site mitigatory action.

The above refers to elements of a documented emergency plan. The guidance on **COMAH** (L111) and especially the HSE guide on emergency planning (HSG191) contain a good deal about the practical considerations in the working of the plan. The following items need consideration:

- Identification of the major incident risks on the site (information on the types of chemicals, their effects and properties together with likely release scenarios).
- Input from relevant external agencies (fire service, utilities, etc.) in developing a workable plan.
- Part of the equipment and resources mentioned earlier is the provision of an **Emergency Control Centre (ECC)**. This is often a dedicated building/room in a relatively safe location (i.e. far enough away from the likely starting point of a major incident). As the name implies, this is the focal point for the emergency operations. The ECC should be kitted out with ready access to such things as site plans, contact information, chemical information and, of course, the necessary communication equipment (radios, telephones).

Co-operation with external agencies

- Specific individuals should have clearly defined roles and responsibilities.
  A hierarchical incident management structure (Main Controller, Incident Controller/Officer, etc.) should be adopted; this is the model used by the emergency services. Though the roles are clearly defined, it may not always be the same individual assigned to this task. Rather there may be a pool of individuals who are available on a rotating 'duty manager' basis to provide adequate cover. There should, of course, also be out-of-hours call-out arrangements.
- Provision of the necessary equipment - this will include communication equipment (e.g. radios, in addition to those used in the ECC), spill containment (absorbents, diking equipment), necessary PPE (such as gas-tight chemical suits and breathing apparatus).

- Maintenance of all the emergency equipment and facilities - these should all be in a state of readiness. The nature of emergencies is that they can happen at any time. So, plan for back-up - have spares available in case of failures, make sure you replenish stocks that get used (absorbents, air from self-contained breathing apparatus, etc.).
- Training of emergency personnel and practising the plan. People are naturally uncertain and confused early in an emergency where there may be little information and lots of things happening. This is especially so when the plan and/or equipment is unfamiliar; uncertainty can allow an emergency to spiral out of control. The plan should not just be a written document which is only ever brought out in a (hopefully rare) emergency. Personnel should be trained so that they are familiar with the procedures and their part in them. This will involve practising the plan on a regular basis.
- Regular testing and review of the plans (at least every three years) and especially if there are significant changes which might affect them. Clearly, a good opportunity to review is after a practice when feedback from participants in a debrief can identify significant deficiencies in workability (e.g. over-complicated communications or command structure causing confusion).
- For extended emergencies, arrangements for welfare facilities should be considered (including outside catering) and also relief of staff who were first on the scene. A major incident can be physically demanding for emergency responders, but it can also be extremely draining mentally and emotionally for decision-makers higher up the chain of command.
- Appointing and training people with the specific responsibility of managing the press.

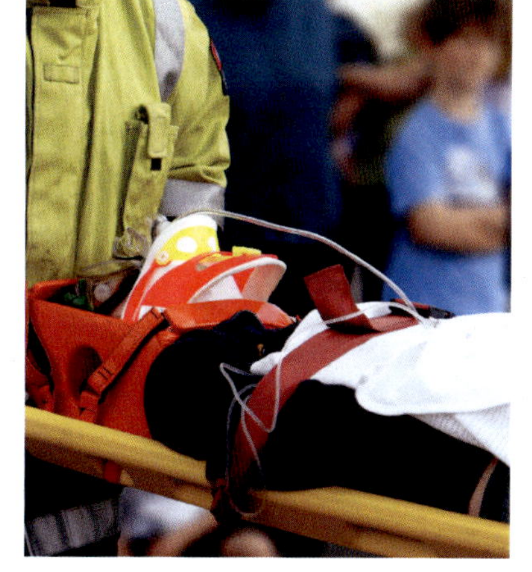

Rescue arrangements within the emergency plan

- Business recovery/continuity issues - a major incident can destroy a factory. Many businesses do not recover after a major fire. Depending on the business risk, a contingency plan, e.g. to transfer production temporarily to another site, may also be needed.

Off-Site

Part 2 of Schedule 4 to **COMAH** identifies the information that should be included in an **external emergency plan (off-site)**:

- The names or positions of persons authorised to set emergency procedures in motion and of persons authorised to take charge of and co-ordinate off-site action.
- Arrangements for:
  - Receiving early warning of incidents, and alert and call-out procedures.
  - Co-ordinating resources necessary to implement the off-site emergency plan.
  - Providing assistance with on-site mitigatory action.
  - Off-site mitigatory action.
  - Providing the public with specific information relating to the accident and the behaviour which they should adopt.
  - The provision of information to the emergency services of other European member states in the event of a major accident with possible trans-boundary consequences.

To enable local authority emergency planners to draw up the off-site plan, the following key pieces of information are required from the operator's safety report (which amounts to an assessment of the major incident risks):

- The hazards that are present:
  - What are the possible sources of harm?
  - What are the dangerous substances, the quantities, their location, their properties and the potential effects on people and the environment?
  - If feasible, what are the reactions of chemicals released and their behaviour in fire?
- What events could give rise to a release of dangerous substances?
- What would be the possible consequences of a release of dangerous substances to people and to the environment?
- Over what distances can these hazards create harmful effects?
- What are the chances of these consequences occurring?

MORE...

The HSE document, L111 *The Control of Major Accident Hazards Regulations 2015 - Guidance on regulations* is available at:

www.hse.gov.uk/pubns/priced/l111.pdf

HSG191 *Emergency planning for major accidents* is available at:

www.hse.gov.uk/pubns/books/hsg191.htm

Reducing the Impact on the Organisation

Emergency plans should not only help the organisation to deal with the emergency as it occurs; they should also have the capacity to help the organisation recover from the impact of the incident on the organisation, including **post-incident recovery**.

Pre-planning for the aftermath and post-incident recovery should allow for a collective understanding of the necessary recovery elements and critical business unit processes so a recovery plan should encompass restoration of the site, infrastructure, and operations.

Disaster recovery operations should prioritise timely and accurate communication to stakeholders, such as Facility Managers, Operations Directors, Emergency Response Teams, Contractors, and if applicable, the public, in order to accelerate recovery without duplicating effort.

Specific guidelines should be written to provide procedures to help facilitate an organised return to normal operating conditions and the process of scaling down resources in an efficient and timely manner should also provide cost benefits to the organisation.

Issues to consider include:

- The on-scene incident commander should approve the release of resources and assign personnel to identify surplus resources based on how well the organisation is coping with the incident.
- A hazardous materials or waste Disposal Plan (part of the Emergency Response Plan) should be implemented.

- Equipment repair, decontamination, maintenance and inspection should be initiated (also part of the Emergency Response Plan). Operators should:
  - Assess damage.
  - Ensure workers can safely access any required facilities.
  - Identify internal and external recovery team contacts and contractors.
  - Identify the scope of work for repair.
  - Develop site specific plans and schedules for executing repairs.
  - Restore operations if it is safe to do so.
  - Institute mitigation measures.
  - Identify 'lessons learned' through post-incident reviews.

Impact assessments and post-incident reviews should be carried out on all affected personnel, equipment, plant and materials and notwithstanding the above, responders must maintain heightened safety awareness until the emergency is declared over.

Recovery planning should also include:

- Co-ordination development, training, and exercise of the disaster recovery plan.
- Establishing and maintenance of a Business Continuity Plan with contacts and networks for recovery of resources and support systems.

Business Continuity Plan objectives should include:

- key operations and critical activities,
- critical processes and strategies for recovery,
- resources to assess, declare, and recover from disruption,
- evacuation and relocation policies and information, and
- key response personnel.

Post-Incident Reviews

The primary purpose of post-incident reviews is to identify deficiencies in the response plan and determine necessary actions to correct the deficiencies. The post-incident reviews can often reveal which response procedures, equipment, and techniques were effective, and which were not and the reason(s) why. These reviews can lead to 'lessons learned' and should be reflected in the response plan, training efforts, and exercise objectives.

Ongoing Monitoring and Maintenance of Emergency Plans

Emergency plans need to be kept up to date to take account of changes and this should be carried out on a planned, regular basis, or when significant changes occur (increased inventory, new hazards, etc.). A good time to appraise current plans is during a rehearsal/test debrief. No plan is ever perfect. Testing often reveals problems with the plans, such as uncertainty of roles or poor communications.

Review is a process which examines the adequacy and effectiveness of the components of the emergency plan and how they function together. The review process must take into account:

- All material changes in the activity.
- Any changes in the emergency services relevant to the operation of the plan.

- Advances in technical knowledge.
- Knowledge gain as a result of major accidents either on site or elsewhere.
- Lessons learned during testing of emergency plans.

Additionally, **COMAH** requires that a review of the adequacy and accuracy of the emergency planning arrangements should follow any modifications or other significant changes to the establishment. Under these circumstances, operators should not wait until the three-year review is due to review their emergency plans.

Role of External Emergency Services and Local Authorities in Emergency Planning and Control

The **COMAH Regulations** require the local authority to consult with those people and organisations that will have a role in the off-site emergency response arrangements. They have a responsibility to consult with the emergency services in the preparation of the off-site emergency plan, so their concerns and recommendations are taken into account in developing and resourcing the plan.

The relevant **environmental agency** is normally consulted early and regularly during the development of the off-site plan to dovetail its response with that of the local authority. The CA should be frequently consulted to assess the regulatory sufficiency of the off-site plan.

The local authority has to consult the appropriate **health authorities** as they have a responsibility to contribute to safeguarding the public health of the population within their geographical area. Thus, they need to be aware of potential major accident risks, in order to dovetail their emergency plans and health service arrangements with those of the emergency services and local authority.

It may also be necessary to consult other organisations in addition to those specifically identified by the Regulations who might become involved and whose roles would need to be included in the off-site emergency plan. These organisations may include the Department for Environment, Food and Rural Affairs (DEFRA), the Food Standards Agency (FSA), the National Assembly for Wales and the Scottish Executive Rural Affairs Department, and local **water/gas/electric companies/authorities**.

Where there is reliance on mutual aid agreements such as Fire, Medical and Security to support the off-site emergency response, the details need to be included in the off-site emergency plan and there should be sufficient consultation to ensure that those involved are aware of their roles and responsibilities.

To ensure that the off-site plan dovetails with the on-site emergency plan, the local authority will need to consult closely with the operator on both the on-site response arrangements and the interfaces between the on- and off-site plans.

The local authority must also consult the **public** when preparing the off-site emergency plan. This could include: consultation with **elected councillors** at county, borough or parish level (or equivalents); or consultation with specially established **groups representing residents** in the vicinity of the site.

Elected councillors will be able to use appropriate channels of communication with the public in the vicinity of the major hazard establishment to obtain their views on the developing emergency plan.

In order that such a plan operates smoothly and efficiently, it is important that responsibilities are set down and understood. This will include non-company personnel, as external services such as the following are likely to be involved in both the development and implementation of the plan:

- Police.
- Fire.

- Ambulance.
- Welfare.
- HSE.
- Local companies.
- Environment Agency.
- DEFRA.
- Technical expertise.
- Electricity company.
- Gas company.
- Water company.
- Local transport.

Someone within the company should also be trained in the responsibility for dealing with the media, as this can have a profound effect on company image.

STUDY QUESTIONS

14. Draw up an emergency plan for dealing with victims of an explosion in a factory manufacturing paints and varnishes.
15. As safety practitioner for a large organisation with premises on a number of different sites, you are responsible for the emergency evacuation procedures. Explain how you would:
    (a) Develop and plan these procedures.
    (b) Ensure that all employees know the procedures.
    (c) Monitor the effectiveness of the procedures.

(Suggested Answers are at the end.)

Summary

Industrial Chemical Processes

We have:

- Considered how temperature, pressure and catalysts can impact on the rate of reaction.
- Examined the heat of reaction in terms of exothermic reactions - those emitting heat, and also the exothermic runaway reaction.
- Discussed examples of exothermic reaction (i.e. combustion) and examples of runaway reactions including the Bhopal incident in 1984.
- Identified ways of controlling exothermic and runaway reactions.

Storage, Handling and Transport of Dangerous Substances

We have:

- Examined what should be considered in a risk assessment of dangerous substances (with reference to **DSEAR** Regulation 5).
- Discussed the way in which dangerous substances should be stored safely, with particular reference to:
  - Storage methods and quantities - bulk storage, intermediate storage, drum storage and specific locations.
  - Storage of incompatible materials and segregation requirements.
  - Containment of leaks or spillages including bunding and problems encountered during filling and transfer.
- Identified the main principles for the safe storage and handling of chemicals with reference to:
  - Flow through pipes.
  - Filling and emptying containers.
  - Dispensing, spraying and disposal of flammable liquids.
  - Dangers of electricity in hazardous areas.
- Identified the main principles associated with the transport of chemicals including:
  - Loading and unloading tankers and tank containers.
  - Labelling vehicles and packaging of substances.
  - The importance of driver training programmes and the role of the Dangerous Goods Safety Adviser.

Hazardous Environments

We have:

- Outlined the principles of resistance to mechanical damage, protection against solid objects and dusts and protection against liquids and gases for electrical equipment used in hazardous environments.
- Discussed the use of electrical equipment used in wet environments including corrosion and degradation of installation and damage to the equipment.
- Considered the classification of hazardous areas using zoning.
- Outlined the use of permits to work.

- Discussed the principles of pressurising and purging with regard to electrical equipment used in hazardous environments.
- Identified the different types of equipment suitable for use in hazardous environments:
  - Intrinsically safe equipment.
  - Flameproof equipment.
  - Type 'e' equipment.
  - Type 'N' equipment.
  - Pressurised apparatus 'p' - using pressurisation and purging.

Emergency Planning

We have:

- Identified the need for emergency preparedness in an organisation, with reference to relevant legislation.
- Considered the ways in which preparation can ensure consequences of an emergency situation can be minimised, including planning for:
  - First-aid/medical provision.
  - Fire evacuation.
  - Spill containment.
- Outlined the development and maintenance of emergency plans, including:
  - The content of both on-site and off-site plans, for major emergency scenarios in order to meet regulatory requirements.
  - How to reduce the impact of the emergency on the organisation, including post-incident recovery.
  - The need for ongoing monitoring and maintenance of emergency plans.
- Examined the role of external emergency services and local authorities in emergency planning and control.

Learning Outcome 11.6

NEBOSH National Diploma for Occupational Health and Safety Management Professionals

ASSESSMENT CRITERIA

- Summarise what needs to be considered during maintenance, inspection and testing of work equipment and machinery.

LEARNING OBJECTIVES

Once you've studied this Learning Outcome, you should be able to:

- Explain safe working procedures for the maintenance, inspection and testing of work equipment according to the risks posed.
- Explain the principles of control associated with the maintenance of general workplace machinery.

Contents

Work Equipment and Machinery Maintenance

IN THIS SECTION...

- Maintenance of work equipment can expose those carrying out the work to a range of different hazards. Key precautions include risk assessment to enable suitable control measures to be put in place, the use of trained, competent maintenance workers, measures such as permit to work, ensuring suitable means of access, and physical isolation of the equipment.
- Statutory duties for the maintenance of work equipment are included in:
  - **Provision and Use of Work Equipment Regulations 1998 (PUWER)**.
  - **Lifting Operations and Lifting Equipment Regulations 1998 (LOLER)**.
  - **Control of Substances Hazardous to Health Regulations 2002 (COSHH)**.
  - **Pressure Systems Safety Regulations 2000 (PSSR)**.
  - **Carriage of Dangerous Goods and Use of Transportable Pressure Equipment Regulations 2009**.
  - **Personal Protective Equipment at Work Regulations 1992**.
  - **Electricity at Work Regulations 1989**.
- The three principal strategies for ensuring well-maintained equipment are:
  - Planned preventive maintenance.
  - Condition-based maintenance.
  - Breakdown maintenance.
- Factors to consider in developing a planned maintenance programme for safety-critical components include:
  - Importance in the process.
  - Machine complexity.
  - Relationship with other machines.
  - Availability of replacement equipment.
  - Identification of critical components.
  - Environmental factors.
  - Maintenance capability within the company.
  - Maintainability within the design of the machine.
- **PUWER** requires work equipment to be inspected at appropriate times or suitable intervals to ensure that health and safety conditions are maintained and that any deterioration can be detected and remedied in good time. The factors to be considered in determining inspection regimes include:
  - The type of equipment.
  - Where it is used.
  - How it is used.
- As part of an inspection, a functional or other test may be necessary to check that the safety-related parts are working as they should be and that the work equipment and relevant parts are structurally sound.

- Setting, cleaning and maintaining machinery are all activities outside the normal operation of the machine which can expose persons involved to a greater degree of risk. There is a requirement for maintenance operations on machinery to be carried out without exposing the persons involved to risks to their health or safety by employing:
  - Safe systems of work.
  - Permits to work.
  - Isolation.
  - Procedures for working at unguarded machinery.
- During repair, service or maintenance work it is often necessary to isolate machinery from potential uncontrolled energy sources.
- Work equipment can mechanically fail as a result of excessive stress, abnormal external loading, metal fatigue, ductile failure, brittle failure, buckling and corrosive failure.
- The aim of Non-Destructive Testing (NDT) is to test without having to destroy the integrity of the material or component. A number of NDT techniques are available, each with its advantages and limitations: dye penetrant, acoustic emission, ultrasonic, radiography (gamma and X-ray), eddy current and Magnetic Particle Inspection (MPI).

Hazards and Control Measures Associated with the Maintenance of Work Equipment and Machinery

Legal Requirements

Maintaining machinery is an activity that is outside the normal operation of the machine which can expose persons involved to a greater degree of risk. The requirement for maintenance operations on machinery to be carried out without exposing the persons involved to risks to their health or safety is contained in Regulation 22 of the **Provision and Use of Work Equipment Regulations 1998** (**PUWER**).

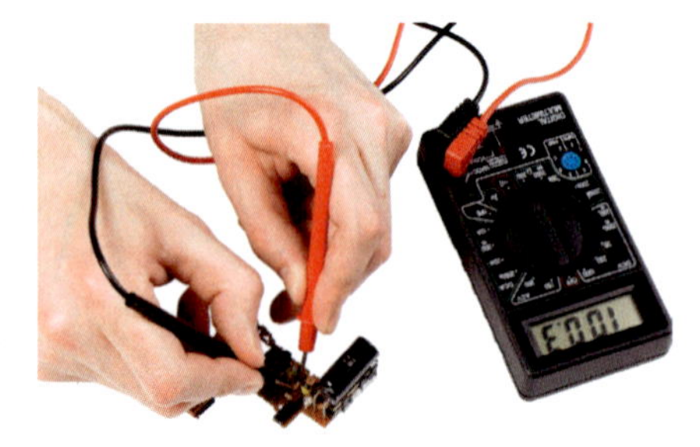

Maintenance of work equipment

Hazards

Hazards that maintenance technicians and engineers frequently encounter might include:

- **Hazardous substances**, such as: fuels, oils and greases, paints, cleaning solvents and acids. Exposure to fibrous dust is also a common hazard when repairing or replacing friction linings such as those found on brakes and clutch mechanisms.
- Due to the nature of the work, it is often necessary to **remove the safeguarding** to gain access to the moving parts of the machine requiring attention, which will mean exposing workers to the dangerous parts of the machinery. Inadvertent start-up of the machine during maintenance work is an ever-present danger.
- Maintenance workers may be required to carry out work in places where it is **not certain what the hazards will be**. A failed or broken down piece of machinery may not have failed to safety and there is always the possibility of stored pressure or energy that may injure maintenance engineers. Equipment failure may result in sharp shards of metal from broken parts and these can cause deep cuts which can become infected after being contaminated with dirt, oil or grease. It is also possible for parts of failed equipment to collapse or fall without warning, resulting in crush injuries and broken bones.

- It is often necessary to carry out work in areas with restricted space where there is poor access, limited lighting and little or no ventilation. Maintenance work is often carried out in **unpleasant or dirty working conditions** and, due to the nature of the work, repairs have to be undertaken in cold, wet, muddy conditions, particularly on construction sites, and this can increase the risk of injury.
- Repair and maintenance is often carried out on pumps and drains in culverts and pits, where rodent urine and faeces is a common hazard resulting in exposure to bacteria such as leptospirosis and *e. coli*. Engineers can also be exposed to HIV and other bacteria and viruses from human waste on sewage farms and in pumping stations.
- The **repairs required can also increase the risks**, particularly where it is necessary to carry out hot work, such as cutting, welding, grinding and burning, with the associated hazards of fire, noise, glare, heat, and possibly the depletion of oxygen. The use of electrical tools also increases the risk of injury, especially when work is carried out in wet or inclement conditions or in flammable atmospheres, increasing the risk of electrocution or burns.
- **Carrying out running adjustments and checks following repairs** often requires the equipment to be energised and this presents additional hazards, particularly where the checks are carried out with safeguarding removed.

From this limited list it is apparent that the risks associated with maintenance work can be higher than normal day-to-day operation of the equipment.

To reduce the risk of injury while maintenance work is being carried out, it is necessary to undertake a thorough risk assessment to enable suitable control measures to be put in place.

Control Measures

Employers should ensure that:

- Work equipment is designed so that maintenance operations involving a risk to health or safety can be carried out while the work equipment is shut down.
- Maintenance operations can be carried out without exposing the person carrying them out to a risk to their health or safety.
- Measures can be taken for the protection of any person carrying out maintenance operations which involve a risk to his health or safety.

If equipment has to be running or working during a maintenance operation and this presents risks, measures should be taken to enable the operation of the equipment in a way that reduces the risk. These measures include further safeguards or functions designed into the equipment, such as limiting the power, speed or range of movement that is available to dangerous parts or providing protection during maintenance operations. Examples are:

- Providing temporary guards and barriers.
- Limited movement controls.
- Crawl speed operated by hold-to-run controls.

Other measures that can be taken to protect against any residual risk include wearing PPE and provision of instruction and supervision.

Access to danger zones should be minimised by locating maintenance, lubrication and setting points outside the danger zones. This is to include access to such places, e.g. stairs and ladders with protection against falls as required.

It should be possible to carry out adjustment, maintenance, repair, cleaning and servicing operations with the machinery at standstill. If this is not possible for technical reasons, then these activities must be able to be performed without risk.

Maintenance of work equipment and machinery can expose those carrying out the work to a number of different hazards. Due to the number of different tasks, ranging from repairing and replacing broken items to cleaning and painting, it is important that a thorough assessment of the hazards, risks and control measures involved in the work to be carried out is made prior to the work commencing.

During regular and reactive maintenance tasks, maintenance workers also face risks that may lead to severe injury or illness, such as: trapping extremities by moving parts of machinery; struck by falling, moving or swinging objects; contact with sharp objects; falls from height; slips, trips and falls; exposure to hazardous substances; fire or explosion; electric shock or electric arc; exposure to dust, fume, vapour, gases, or biological agents.

Statutory Duties for Maintenance of Work Equipment

The law distinguishes between maintenance, inspection and "thorough examination". The last one is often also taken to include testing. (We will discuss the types of maintenance later.) "Inspection" is what is done in between thorough examinations (discussed later) and generally involves visual checks (and possibly tests); it can usually be done by the operator/ user.

Inspection of work equipment

There are a number of pieces of legislation that require inspections to be undertaken and recorded against particular types of equipment; a selection of which are summarised below. The requirements can be complex and detailed. You should therefore read the full text of the references indicated:

- **Provision and Use of Work Equipment Regulations 1998 (PUWER)**

  PUWER is the most important piece of legislation regarding maintenance and inspection requirements for machinery and work equipment. Specific, applicable regulations include:

  - Regulation 5, 22 - Maintenance of work equipment in general.
  - Regulation 6 - Inspection (except power presses).
  - Regulation 32 - Inspection of power press guards.
- **Lifting Operations and Lifting Equipment Regulations 1998 (LOLER)**
  - Regulation 9 - Inspection of lifting equipment.
- **Control of Substances Hazardous to Health Regulations 2002 (COSHH)**
  - Regulation 9 - Maintenance of control measures (including local exhaust ventilation).
- **Pressure Systems Safety Regulations 2000 (PSSR)**
  - Regulation 12 - Maintenance of pressure systems.
- **Carriage of Dangerous Goods and Use of Transportable Pressure Equipment Regulations 2009**
  - Various regulations for maintenance and inspection of tanks and also transportable pressure equipment.
- **Personal Protective Equipment at Work Regulations 1992**
  - Regulation 7 - Maintenance of PPE.

Maintenance is required for lifting equipment

- **Electricity at Work Regulations 1989**
  - Regulation 4 - Maintenance of electrical systems.

Hired Work Equipment

PUWER requires work equipment to be:

- Suitable for the purpose for which it is to be used.
- Maintained in an efficient state.
- Inspected as necessary.

Hired or leased equipment must also comply with these requirements and organisations hiring or leasing equipment will need to be assured that companies supplying such equipment are doing so in accordance with the Regulations.

Arrangements (policies) will need to be in place to ensure that risk assessments adequately cover leased and loaned work equipment. Only reputable hire outlet shops should be used who can demonstrate appropriate inspection and maintenance standards, and hired work equipment should be maintained while in use.

Maintenance Strategies

In order to ensure work equipment does not deteriorate to the extent that it may put people at risk, employers, the self-employed and others in control of work equipment are required by **PUWER** to keep it "maintained in an efficient state, in efficient order and in good repair".

Effective maintenance programmes not only help in meeting **PUWER** requirements but can also serve other business objectives, such as improved productivity and reduced environmental impact.

The frequency and nature of maintenance should be determined through risk assessment, taking full account of:

- The manufacturer's recommendations.
- The intensity of use.
- The operating environment (e.g. the effect of temperature, corrosion, weathering).
- User knowledge and experience.
- The risk to health and safety from any foreseeable failure or malfunction.

Safety-critical parts of work equipment may need a higher and more frequent level of attention than other aspects, which can be reflected within any maintenance programme. Breakdown maintenance, undertaken only after faults or failures have occurred, will not be suitable where significant risk will arise from the continued use of the work equipment.

There are three main strategies for maintenance of work equipment which we will consider in more detail below.

Planned Preventive Maintenance

The basis of routine maintenance is that equipment is inspected and vulnerable parts are replaced at regular intervals or after a certain number of hours of use. How frequently should this maintenance be done? If it is done too frequently, the equipment will run effectively but the maintenance costs will be very high. If it is done too infrequently, the maintenance cost will be low but the system will be subject to failures.

If breakdowns are to be avoided, some form of planned maintenance is essential. Such a system allows machine servicing to be planned to minimise lost production time.

Planned maintenance is based on the fact that any machine or plant item consists of various components, each of which has a definable working life. These can be classified into one of the following four item types:

Planned preventive maintenance in progress

- **Maintainable**: a maintainable item is one that will last for many years if serviced regularly, e.g. an item that requires regular lubrication (such as a gearbox).
- **Replaceable**: a replaceable item is one with a limited life which should be replaced at regular intervals if sudden and possibly disastrous failures are to be avoided, e.g. a drive belt.
- **Inspectable**: an inspectable item also has a limited life, but potential failure can be identified by some form of measurement, e.g. play in a bearing or wear on electrical contacts.
- **Long-life**: a long-life item is one that is likely to last for many years without failure because it is not subject to any appreciable wear, vibration or stress; e.g. a machine baseplate.

Preventive maintenance is regular maintenance carried out in order to reduce the probability of problems occurring; such work should be timed to occur when operations will not be disrupted. Preventive maintenance has long been recognised as extremely important in the reduction of maintenance costs and improvement of equipment reliability. In practice, it takes many forms. Two major factors that should control the extent of a preventive programme are the:

- Cost of the programme compared with the carefully measured reduction in total repair costs and improved equipment performance.
- Percentage utilisation of the equipment maintained.

If the cost of preparing for a preventive maintenance inspection is essentially the same as the cost of repair after a failure, accompanied by preventive inspections, the justification is small. If, on the other hand, breakdown could result in severe damage to the equipment and a far more costly repair, a scheduled inspection time should be considered.

Ideally, you should take advantage of a breakdown in some component of the line to perform vital inspections and replacements, which can be accomplished in about the same time as the primary repair. This would ensure preventive maintenance is carried out without discontinuing operations unnecessarily.

TOPIC FOCUS

The advantages and disadvantages of the operation of a **planned preventive maintenance system** include:

Advantages

- Safer operation.
- Reduced downtime and lost production.
- Increased reliability of equipment.
- Failure-rate predictions.
- Rapid turnaround of parts.
- Maintenance carried out at times of least disruption.
- Reduced risk from planned maintenance operations.

Disadvantages

- Over-maintenance, with reduced efficiency.
- High costs of parts still with useful life.
- Management time for planning and operating schedules.
- Over-familiarity with tasks leading to complacency.
- Higher skill level.
- Increased storage requirements for spare parts.

The benefits of planned preventive maintenance include:

- **Legal** - to demonstrate that the employer has taken reasonably practicable steps to meet their legal obligations to maintain safe plant and a safe place of work.
- **Business** - an effective maintenance strategy will improve both reliability and the image of the company with customers. Well-maintained work equipment will also be more likely to produce output goods within the tolerances required by the contract, e.g. worn dies on a moulding machine will produce items which, being outside the tolerances, must be rejected. Well-maintained work equipment reduces the chance of an accident, which in turn reduces the potential for injuries and subsequent cost implications. With a regime which ensures that repairs are dealt with efficiently and effectively, staff become less likely to use substandard equipment and are more likely to report damage or defects. This improves the safety culture of the company and will reflect in areas outside those relating to machinery, in that the general attitude to all matters relating to safety is improved.

From all this, it is clear that positive returns are available from effective maintenance. While meeting legal requirements is mandatory, the choice of an appropriate maintenance strategy is founded in positive business principles.

Condition-Based Maintenance

Condition-based maintenance relies on monitoring the condition of safety-critical parts and carrying out maintenance whenever necessary to avoid failure. Maintenance is performed after indicators show that equipment is going to fail or that equipment performance is deteriorating.

Condition-based maintenance should improve system reliability, since maintenance is carried out before failure. It should also decrease maintenance costs, avoiding both unnecessary and breakdown maintenance. Since fewer operations are carried out, there should also be less chance of errors occurring during maintenance.

However, the primary disadvantage is cost arising from setting the system up to monitor parameters such as vibration, temperature, and also non-destructive testing techniques. In addition, unpredictable maintenance periods can cause costs to be divided unequally and the system will increase the number of parts that need maintenance and checking.

Condition-based monitoring involves routine checking of the condition of safety-critical parts. For example, whenever a car is serviced, the remaining thickness of brake linings/shoes and tyre tread depth will be monitored.

Where safety-critical parts could fail and cause the equipment, guards or other protective devices to fail, and lead to immediate or hidden potential risks, a formal system of condition-based maintenance is likely to be necessary.

Breakdown Maintenance

When equipment failure does not have a major effect on production or safety and may be tolerated until repair, then the positive decision to use this as an option can be valid. The strength of this system is that it has a minimal cost in relation to a maintenance resource. The weakness is that the equipment is out of use during a possibly extended breakdown period and that repair has to be arranged. The effect on the equipment may be such that, depending on what has failed, the damage may reduce the effective working life.

Breakdown maintenance

Factors to be Considered in Developing a Planned Maintenance Programme

Whatever the chosen maintenance strategy, the organisation needs to consider a number of factors that may, at times, be in conflict with each other. However, weighing up the factors will help management decide the best strategy to choose. The factors outlined below are 'strategic' factors, i.e. top level, which have to be identified and assessed in relation to the role of the machinery under consideration:

- **Importance in the Process**

  Any strategy adopted should reflect whether the equipment has a critical and central role. For example, if all production depends on a conveyor moving components from one stage of production to the next and if without it, production stops, the maintenance strategy must ensure the minimum downtime during the production process.

- **Machine Complexity**

  Any piece of machinery will have a number of components, any of which may fail. Some are more difficult to gain access to than others and the potential time delay in repair must be considered. Certain components may be so inaccessible that replacement or repair requires total strip-down of the machine. Consequently, some maintenance strategies available may not be an appropriate choice. An example is the replacement of an armature within an electric motor, where the amount of stripping-down justifies that the bearings, etc., are checked at the same time.

- **Relationship with Other Machines**

  In a production line, failure of one machine may cause total failure of the entire line. The position and effect of a possible failure has to be considered, such that the best option for the individual machine may not be the best option for the overall production line. In such cases, the production line should be regarded as a single entity in the first instance, the individual machines being considered separately when reviewing implementation within the chosen maintenance strategy.

- **Availability of Replacement Equipment**

  Maintenance strategy selection is varied when it is possible to replace failed equipment with functional equipment held 'in store' for such an occasion. The availability of the replacement, and the work involved in fitting it and getting it operational, are all factors which must be taken into account, along with the availability of maintenance capability to undertake the work, as detailed below. This factor is an option frequently used for small machines with low purchase cost where it is possible to have replacements available. For example, on construction sites, 110v electric hand tools are usually available as replacements (either on site or with quick delivery) for failed units, the unit normally being returned to a workshop for repair.

- **Identification of Critical Components**

  Within any machine there are a number of key components which will have a major effect on the failure mode of the machine. The identification of such components is important when deciding on the maintenance strategy to be adopted. If studies, for example, show that in a particular make of pump it is the armature which fails regularly, then the strategy chosen must target that component and the maintenance strategy of the other components should be considered in relation to the strategy chosen for the armature; such as full pump maintenance at the time of the armature being attended to, rather than at a lesser frequency which the other components would normally dictate.

Planned preventive maintenance for safety-critical components

- **Environmental Factors**

  Machine characteristics can be considerably altered by the environment in which the machine is required to work. Environmental factors can include: heat, cold, dampness, dust, vibration and vapours. Electrical power supply reliability should be examined as power fluctuations may influence machine operation.

- **Maintenance Capability Within the Company**

  Since any maintenance strategy has to be implemented, the resource requirements in time and skill have to be taken into account. It is inappropriate to consider a strategy if the resources are not available to implement it.

- **Maintainability Within the Design of the Machine**

  The manufacturer has to make provision for maintainability to comply with BS EN ISO 12100:2010, which requires that the following factors are taken into account:

  - Accessibility, taking account of the environment, the dimensions of the human body, and working clothes and tools.
  - Ease of handling and human capabilities.
  - Limitations on the number of special tools and equipment.

- **Unknown Factors**

  People generally believe that they are aware of what factors affect equipment. In practice, many rely on information from records or by discussion, making the information second or third hand. In addition, the reliability of this information, e.g. consistency of completion of records or reporting of failures, must be assessed to ensure that a strategy decision is being made on accurate information. This is the most common failing in selection of the most suitable maintenance strategy. Operatives, for example, may find that an electrical hand tool regularly fails due to the power lead being loose in the socket as a result of poor design. It is not normally reported, as the fixing of the machine is carried out by the operator who simply pushes the nuisance lead back in and does not bother with a boring reporting process. Review of the record for failures would not reveal the true downtime of the machine.

Inspection Regimes

There is a clear distinction between maintenance, inspection and 'thorough examination'. The last one is often also taken to include testing. (We will discuss the types of maintenance later.) 'Inspection' is what is carried out in between thorough examinations (discussed later) and generally involves visual checks (and possibly tests); it can usually be completed by the operator/user.

PUWER requires work equipment to be inspected at appropriate times or suitable intervals to ensure that health and safety conditions are maintained and that any deterioration can be detected and remedied in good time. However, inspection is only necessary where there is a significant risk resulting from incorrect installation, deterioration or as a result of exceptional circumstances which could affect the safe operation of the work equipment.

The purpose of an inspection is to identify whether the equipment can be operated, adjusted and maintained safely and to ensure that any deterioration (e.g. defect, damage, wear) can be detected and remedied before it results in unacceptable risks.

The extent of the inspection required will depend on the potential risks from the work equipment and will need to consider:

- The type of equipment:
  - Are there safety-related parts which are necessary for safe operation of the equipment, e.g. overload warning devices and limit switches?
- Where it is used:
  - Is the equipment used in a hostile environment where hot, cold, wet or corrosive conditions may accelerate deterioration?
- How it is used:
  - The nature, frequency and duration of use will determine the extent of wear and tear and the likelihood of deterioration. In addition, equipment regularly dismantled and re-assembled will require inspection to ensure that it has been installed correctly and is safe to operate.

The inspection itself may vary from a simple visual external inspection to a detailed comprehensive inspection, which may include some dismantling and/or testing.

MORE...

Recommended frequencies for some types of work equipment are provided by the Safety Assessment Federation guidance document MLCC05 - *In-Service Inspection Procedures*, available to download from:

www.safed.co.uk/technical-guides/machinery-lift-and-crane

The frequency of inspections for work equipment needs to be determined by risk assessment, using the knowledge and experience of persons who are competent to determine the nature of the inspection, what it should include, how it should be done and when it should be carried out. Experienced, in-house employees such as a department manager or supervisor may be able to do this, although inspection is often outsourced to experts in that particular field. An inspector should know what will need to be inspected to detect damage or faults resulting from wear or environmental deterioration and will also be able to determine whether any tests are needed during the inspection to see if the equipment is working safely or is structurally sound.

In the 21st century there is technology available that can provide inspection services without the need for employees to be placed in danger. Remotely Operated Vehicles (ROV) and drones are just two solutions to the problem of having to place personnel in danger when inspecting plant and equipment.

These days, drone and ROV inspections are performed in almost every industry that requires visual inspections as part of its maintenance procedures and by using them to collect visual data on the condition of equipment, inspectors avoid having to place themselves in dangerous situations and the organisation saves time and money by perhaps not having to build access platforms or go through the drawn out procedure of implementing a risk assessment and permit to work process with all the manpower and emergency procedures required to ensure the safety of the working party.

On larger machines, drones and ROV can negate the requirement to access confined spaces or work at height by inspecting the outside and the inside, and there is often a further saving in machinery and equipment downtime related to the fact that expensive and time-consuming access platforms do not have to be constructed. Drones can inspect immediately and the results are available immediately.

Drone/ROV data also photographs the condition of an asset over time and the company will have a record of the asset's life history that can be accessed at any time.

Need for Functional Testing of Safety-Related Parts

Safety-related parts include:

- Fixed guards.
- Interlocks and protection devices associated with an interlocked guarding system on the dangerous parts of a machine.
- Controls and emergency controls, such as emergency stops, necessary for the general safe operation of the equipment.

Functional testing of safety systems

These safety-related parts should be tested to ensure that they are operationally effective. Circumstances that would require this include:

- Prior to first use of a newly purchased/installed machine.
- After any form of maintenance has been carried out on the machine.
- After an event that may have involved or affected the safety of the machine.
- Routinely during normal use at a frequency dictated by the risk presented by the machine. This would normally be at the start of every shift, but might be more frequently if the machine were subject to heavy usage.
- After any occasion where the machine is set up or installed at a new location where its disassembly, transport and reassembly may have affected the effectiveness of the safety device.

This testing may be done by visual inspection alone or it may form a part of the routine scheduled maintenance of the machine.

So, for example, a dangerous machine such as a power-operated, paper-cutting guillotine should be subjected to:

- Daily operator checks on interlocking or photoelectric guards.
- Monthly operator checks on sweep-away guards.
- Six-monthly inspections of all safety components, such as brakes, clutches, interlocks, switches and cams, carried out by a competent guillotine engineer.

To check that relevant parts are structurally sound, non-destructive testing of safety-critical parts might be required.

The need for any testing should be decided by the competent person who determines the nature of the inspection.

Setting, Cleaning and Maintaining

Setting, cleaning and maintaining machinery are all activities outside the normal operation of the machine which can expose persons involved to a greater degree of risk. The requirement for maintenance operations on machinery to be carried out without exposing the persons involved to risks to their health or safety is contained in Regulation 22 of **PUWER**.

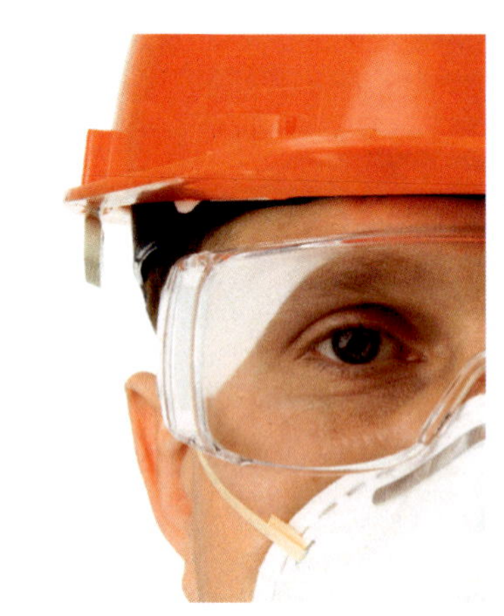

Personal protective equipment

Employers should ensure that:

- Work equipment is designed so that maintenance operations involving a risk to health or safety can be carried out while the **work equipment is shut down**.
- Maintenance operations can be **carried out without exposing the person carrying them out to a risk** to their health or safety.
- **Measures can be taken for the protection of any person** carrying out maintenance operations which involve a risk to their health or safety.

If equipment has to be running or working during a maintenance operation and this presents risks, measures should be taken to enable the operation of the equipment in a way that reduces the risk. These measures include further safeguards or functions designed into the equipment, such as limiting the power, speed or range of movement that is available to dangerous parts or providing protection during maintenance operations. Examples are:

- Providing temporary guards.
- Limited movement controls.
- Crawl speed operated by hold-to-run controls.

Other measures that can be taken to protect against any residual risk include wearing PPE and provision of instruction and supervision.

Access to danger zones should be minimised by locating maintenance, lubrication and setting points **outside** the danger zones. This is to include access to such places, e.g. stairs and ladders with protection against falls as required.

It should be possible to carry out adjustment, maintenance, repair, cleaning and servicing operations with the machinery at a standstill. If this is not possible for technical reasons, then these activities must be able to be performed without risk.

Safe Systems of Work

Most of the topics so far in this Learning Outcome relate to the guarding of machinery in its **working state** (and to some examples of non-functioning state). However, machinery is not always working; tasks are required to set up and adjust the machine, for which the choice of guard becomes important. Likewise, cleaning and maintaining the machinery is important to ensure that it is fit to function when required and to prolong its operational life. The basic premise starts with having in place a safe system of work which manages the risks. It may not be possible to introduce the same level of protection as when the machine is working, but remember that time of exposure should be taken into account.

To be effective, a safe system of work must bring together the following elements (the acronym **'PEME'** is often used as an aide memoire in this case):

- People - who is the SSW for? What level of competence or technical ability should these people have to carry out the work safely?
- Equipment - what equipment is required to complete the task safely? What other equipment (such as PPE) is needed to ensure the safety of workers?
- Materials - what materials will be used or handled during the work? What materials will be required to ensure the task is carried out safely?
- Environment - where will the work take place? Is it hot/cold/dusty or in a flammable atmosphere? Is there good lighting and plenty of space to work?

It is emphasised that situations involving unguarded machinery under power in any phase of its life cycle should be avoided by appropriate design measures wherever technically feasible. Alternatives may include the use of completely different types of machine to achieve the same end product. When there is no alternative, the following general precautions should be observed by properly trained and supervised personnel:

- Safe access, with firm footholds (and handholds where necessary) should be provided. This should be free of obstruction and any material likely to cause slipping.
- Where the hazards include entanglement and drawing-in, loose clothing, neckties, gloves, rings and other jewellery, long hair (unless tied back and/or covered), fabric, first-aid dressings and bandages, and any other material likely to be caught up should be avoided. For any close approach, close-fitting overalls with close-fitting cuffs and no external pockets should be provided. It should be borne in mind that even when guarded against contact, entanglement hazards may be within reach of adjacent loose or stray material, etc. Material in the machine, e.g. material being processed, or by-products such as swarf, may also present an entanglement hazard.
- Where the hazards include impact or penetration due to flying objects, including small particles and dust, appropriate eye protection should be worn.
- Precautions against impact injuries due to kickback are necessary on certain types of cutting and abrasive machinery, particularly where workpieces are manipulated by hand. These include the following:
  - Provision of backstops on vertical spindle moulding work.
  - Ensuring circular saw blades are adjusted to protrude through the material being cut, and that riving knives are of the correct thickness.
  - Ensuring work rests are adjusted close to abrasive wheels or tool rests are correctly adjusted.
  - Ensuring that cutter speeds, or wheel speeds, are correct for the task in question: this includes ensuring that circular saw blades are of large enough diameter to have the correct tooth speed; machines should be labelled with the minimum blade diameter.
- Precautions against impact injuries due to bursting generally involve ensuring that relevant rotating equipment, and any abrasive wheels, etc., used with it are marked clearly with their maximum speeds.

- There are also practices relating to the approach of workers to machinery, which are relevant to most of the types of mechanical hazards we have looked at. These include the following:
  - Limiting closeness of approach, e.g. in work near overhead travelling cranes.
  - Provision and use of manual handling devices, e.g. tongs for forging work, push sticks for circular saws and spindle moulders, or push blocks for planing machines.
  - Provision of jigs and holders for workpieces, e.g. for vertical spindle moulding, or for cutting irregular material on circular saws.
- Emergency stop controls should be readily accessible.

It is not always possible to eliminate hazards or to design completely adequate safeguards to protect people against every hazard, particularly during such phases of machine life as commissioning, setting, process changeover, programming, adjustment, cleaning and maintenance, where often direct access to the hazardous parts of the machine may be necessary.

There are also a number of types of machinery where, at present, it is recognised that complete safeguarding cannot be provided even for operational activities. For some of these types of machinery, safe working practices are specified, e.g. in statutory regulations. It should be emphasised that safety of machinery depends on a combination of hazard minimisation measures, safeguards and safe working practices. These should take account of activities during all phases of the machine's life.

Safe working practices should be taken into account at the design stage, since the provision of jigs, fixtures, fittings, controls and isolation arrangements will be frequently involved.

Permit to Work

Many systems, including those for working on equipment, may involve a number of factors that need to work together for the system to minimise the risk and therefore comply with legal requirements. Some of these factors may include:

- The working environment.
- Tools, plant and equipment, inspection and maintenance systems.
- Operator training.
- The enforcement of rules.
- Supervision.

Documentation will be required for all but the simple systems. Where a greater degree of control is required (because of the higher risks and therefore high potential loss), a formalised method is required to minimise the chance of error.

Effective control may be introduced by means of a written system known as a **permit-to-work system**. This sets out in writing all relevant checks and controls that need to be undertaken before work can commence. The following figure shows an example of a permit.

Machinery and Equipment Certificate of Appointment	Part II	Machinery and Equipment - Certificate of Appointment Unfenced Machinery/Equipment
Part 1 REQUEST FOR CERTIFICATE OF APPOINTMENT to approach unfenced machinery for the purpose of observation which is found immediately necessary.		This certificate appoints Name: .... Badge No. .... Dept: .... to approach Machine/Equipment .... BT NO: ...... Loc: ...... Dept: .... for observation purposes only of the process or part(s) as detailed herewith: ....
Signed: .... (Supervisor		Signed: .... (Authorised Person) Date: ....
ID No. ...... Dept)	Note:	Time: .... am/pm
Darted:	Part III	ONLY PERSONS AUTHORISED BY THE PLANT ENGINEER ARE PERMITTED TO ISSUE THIS CERTIFICATE OF APPOINTMENT
A Certificate of Appointment may only relate to one person. The counterfoil must be handed to the authorised person and retained by him until the permit is returned with Part III completed. Instruction can then be given to clear Part IV and resume normal operation.		(To be completed by person appointed in Part II) I hereby declare that *(1) task designated is complete/incomplete. *(2) all guards replaced or machine equipment is left in a safe condition. * (Delete whichever is not applicable)
	Part IV	Signed: .... (Appointed Person) Date: .... Time: .... am/pm I hereby declare that this certificate is now cancelled. Signed: .... (Appointed Person) Date: .... Time: .... am/pm
Serial No:		THIS CERTIFICATE IS VALID ONLY FOR THE SHIFT IN WHICH IT IS ISSUED OR THE COMPLETION OF THE OBSERVATION (WHICHEVER IS EARLIER). WHEN COMPLETED MAIL BOTH PARTS TO PLANT SAFETY ENGINEERING DEPARTMENT. FORM NO:

(Printed on back of certificate)

PRECAUTIONS TO BE TAKEN FOR SAFE ENTRY INTO TRANSFER MACHINES FOR THE PURPOSE OF OBSERVATION

1. A close fitting single piece overall suit in good repair shall be worn. It shall have no loose ends and no external pockets except a hip pocket. It shall be worn in such a way that it completely covers all loose ends of outer clothing.
2. No guard shall be removed from any part of machinery except when the observation cannot otherwise be carried out and it shall be replaced immediately the observations have been completed.
3. Appointed persons shall make proper use of any appliances provided for the safe carrying out of the observation.
4. Appointed persons shall make proper use of the secure foot-hold and hand-hold where provided as a precaution against slipping.
5. If a ladder is used it shall either be securely fixed, lashed or footed.
6. Another person, who has been instructed as to what to do in case of emergency, shall be immediately available within sight or hearing.
7. Where there is a foreseeable risk of eye injury from the machining process, the appropriate eye protection shall be worn.

SPECIAL NOTES:

- Only persons who have been appointed in writing overleaf, shall enter Transfer Machines specified for the purpose of observation.
- Only persons who have obtained the age of 18 shall be appointed.
- An appointed person must not perform any operation other than that specified in Part II of the certificate.
- The appointed person must have been instructed as to the requirements of the "Procedure for the Safe Entry into Transfer Machines" and be sufficiently trained for the work and be acquainted with the dangers from moving machinery.

Sample permit to work

A key point to note is that the person in charge signs that all protection is in place before the certificate is issued and work can commence. This should (as much as possible) remove the human error element (and the possibility for mistakes) from the maintenance process. A weakness of the permit to work is that if not monitored, checked regularly and enforced, it can quickly fall into disuse with obvious consequences. Any check made on the permit-to-work system should ensure that all the relevant information (but only relevant information) is asked for. Non-relevant information has the effect of making the permit system appear to be only a paperbound exercise and of no practical value.

Isolation

Although safeguards are provided which prevent access during most phases of machine operation, they may not be effective at all times because of the need to gain access to hazardous areas, e.g. for setting-up purposes. Where interlocked guards are provided, many short-term activities such as adjustment or lubrication may be carried out safely by relying on the interlocks.

Isolation measures include locks, clasps, tags, closing and blanking devices, removal of mechanical linkages, blocks, slings, and removal from service.

TOPIC FOCUS

It may be necessary to **isolate** machinery from potential uncontrolled **energy sources** during repair, service or maintenance work.

Energy sources may be in the following form:

- Electrical.
- Mechanical.
- Hydraulic.
- Pneumatic.
- Chemical.
- Thermal.
- Gravitational.
- Radiation.
- Stored or kinetic energy.

Machinery isolation or lock-out is the isolation and safe removal of the energy source(s) from an item of machinery in such a way as to prevent the possibility of inadvertent energising of the machine. Each energy source should be isolated and locked out at each isolation point along the energy source route where practicable.

Isolation of energy sources is considered in more detail below.

Procedures for Working at Unguarded Machinery

Whether guards are in place or not, the hazards imposed by the equipment still exist. Where guards are not in place, these hazards should be managed to prevent them becoming a risk to the operator or others. Key to this is reducing the exposure and thus the risk.

PUWER, Regulation 7, refers to "specific risks", into which category this aspect would come. The controls imposed are that:

- The use of the equipment is restricted to those persons given the task of using the equipment.
- Only those people designated can undertake the work. (Note that generally guarding is designed to protect everyone in the vicinity of the machinery from the danger and there is no restriction on persons approaching correctly guarded equipment. However, in this case, the designation of persons (preferably in writing) is the positive action required to minimise the risk.)
- Persons so nominated are specifically trained and competent to deal with the particular risks involved.

Isolation of Energy Sources

It may be necessary to isolate machinery from potential uncontrolled energy sources during repair, service or maintenance work. Here we outline means for achieving this.

External Isolation and Energy Dissipation

Interlocks provide reliable and secure means of interrupting the power sources but, because of the nature of non-operational activities, the risk levels involved may frequently be higher, and therefore more reliable and more secure means of interruption may be called for. Interruption of the following may be involved:

- **Mechanical power transmission**: by isolating clutches, or by removal of drive belts or chains, or shaft sections. Scotching may also be used.
- **Electrical power**: by isolating switches, by removal of fuses, or by removal of plugs from sockets. Earthing may also be used.
- **Hydraulic or pneumatic power**: by isolating valves, by electrical isolation of hydraulic pumps, or by disconnection from pneumatic mains. Open venting to atmosphere may also be used.
- **Services**: isolation of water, steam, gas or fuel supplies.
- **Process and material supplies**: isolation of process lines and line blinding or blanking-off.

Provision for these facilities has to be made at the design stage.

DEFINITION

SCOTCH

Mechanical scotches are required on certain types of presses to protect the operator when reaching between the platens.

A **scotch** is a mechanical restraint device that can be inserted into the machine so as to prevent physical movement of dangerous parts. These scotches can be linked to the guard operation so they are automatically positioned each time the guard is opened.

It should only be possible to actuate it when the dangerous part (platen of down stroking press or injection moulding machine) has been fully retracted. Alternatively, a progressive scotch will retain the platen at any position of its travel. Scotches can be used to retain a rise and fall guard in the raised position and should be used to prop the body of a tipper lorry when work is being carried out on the vehicle hydraulic system or transmission.

Internal Isolation and Energy Dissipation

In each case, any residual energy storage or material in the machine or equipment may also have to be dealt with, as follows:

- **Mechanical power**: allow flywheels or high-speed rotating parts, e.g. centrifuge bowls, to run down and to minimise the potential energy of other parts.
- **Electrical power**: discharge of capacitors, or disconnection of stand-by batteries.
- **Hydraulic power**: discharge of accumulators, or relaxation of pressurised pipework.
- **Pneumatic power**: discharge of air pressure throughout the system (except where used for hold-up).
- **Services**: residual steam, gas or fuel may need to be vented, purged or drained.
- **Process and material supplies**: emptying, venting, purging, draining and/or cleaning may be required before entry.

Provision for these facilities has to be made at the design stage.

Checking for Absence of Hazards

After external and internal isolation and dissipation of hazards, a check should be made to ensure that no hazard remains. On occasion, this may involve special instruments or sensors, though this is more often the case with non-mechanical hazards.

Security of Safety Measures at the Control Station

Where controls are remote from the plant or machinery, the operator may be unable to see other people, such as maintenance personnel, who may be within the guarded area. In these circumstances, it is recommended that the relevant controls and facilities should be lockable, or such that they can only be operated by the use of keys or tools.

Multi-padlock hasps or a multiple key exchange arrangement should be provided where more than one person is at risk. Each person should apply an individual padlock or key to each relevant control.

Typical Causes of Failures

Abnormal external loading can place **excessive stress** on work equipment and machinery components, which can lead to failure with potential safety implications.

DEFINITIONS

STRESS

The force applied per unit area.

STRAIN

The fractional distortion due to the stress (e.g. amount a wire stretches compared to its original length).

Below certain stress limits, the ratio of stress/strain is found to be constant (Young's modulus), that is, there is a given distortion for each unit increase of stress. The ratio depends on the material and type of stress. There are a number of different categories of stress; the main ones are shown in the following figure:

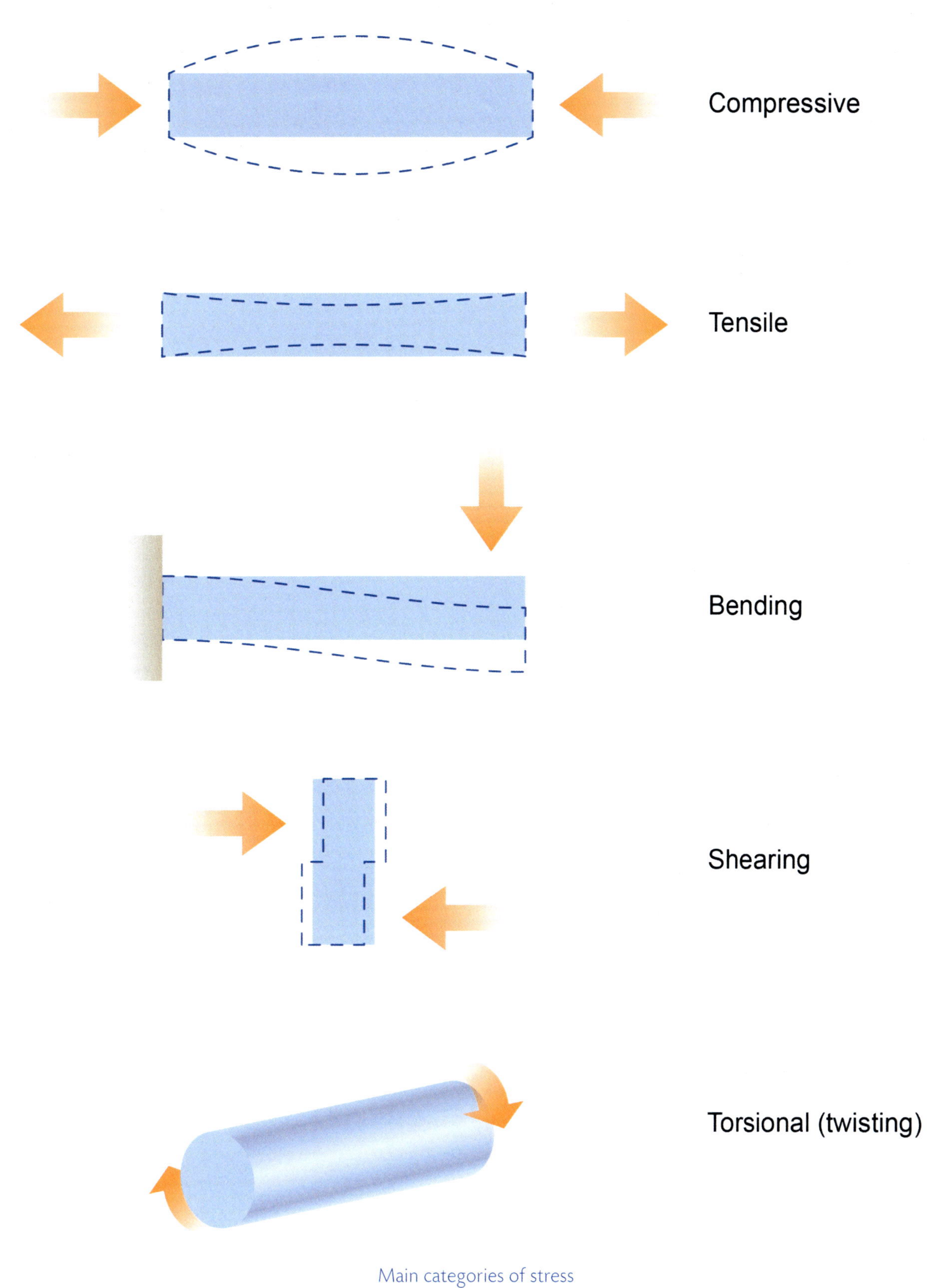

Main categories of stress

If a piece of wire is subjected to increasing tensile stress (e.g. by attaching weights to one end), it will extend:

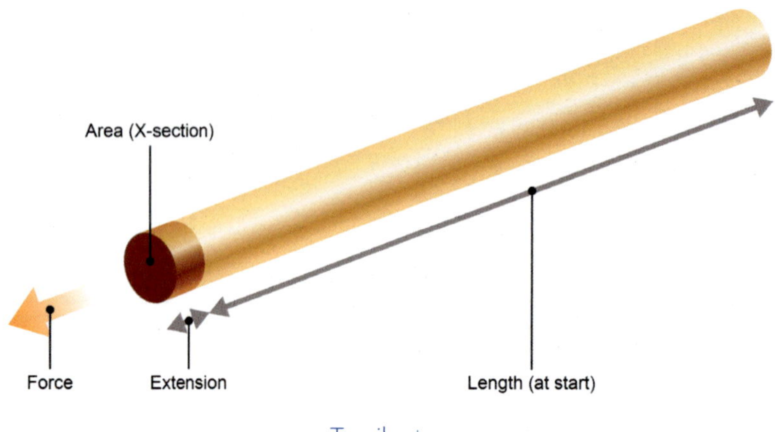

Tensile stress

If we then plot the stress against the measured strain, we see three basic types of behaviour, depending on the material type:

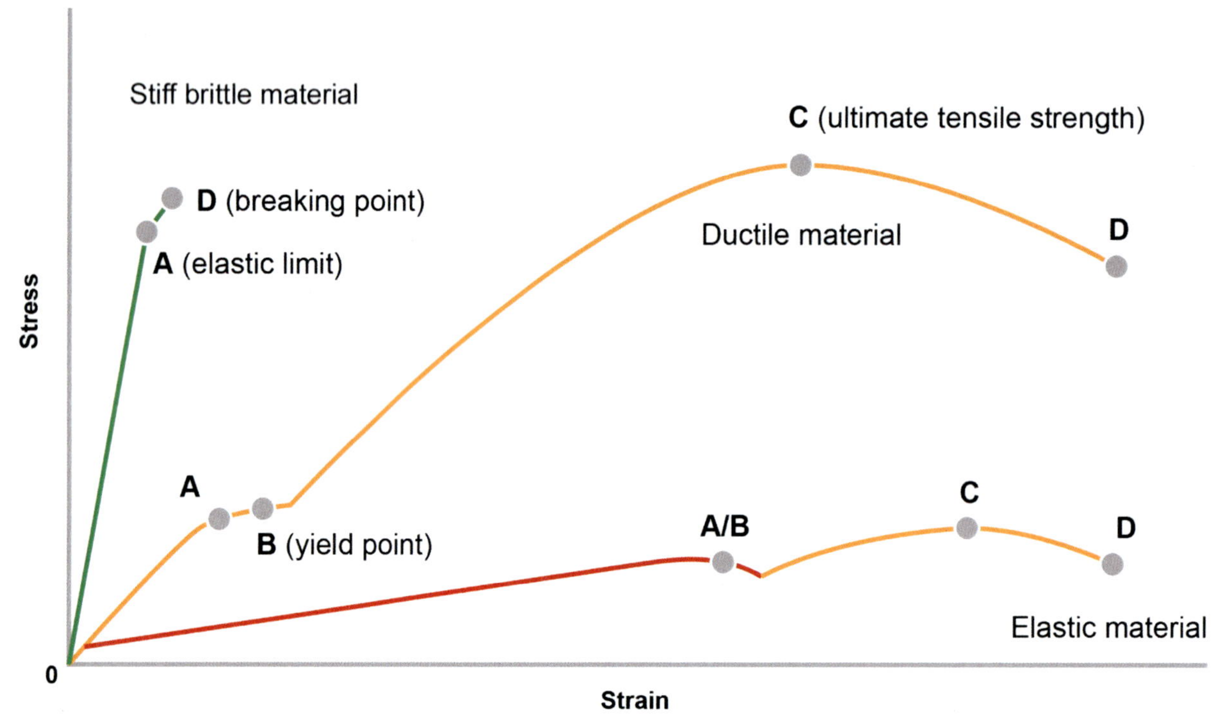

Types of behaviour of different materials

Let us initially focus on the central curve ('Ductile material'). In the aptly-named 'elastic region' (0 - A on the above curves), there is no permanent deformation; the wire goes back to its original dimensions once the stress is removed. The extension per unit load is constant here.

At the yield point (B on the curve), the material begins to plastically deform (i.e. it is no longer elastic). Beyond this, we enter the plastic region where there is a sudden increase in extension.

Ductile materials include mild steel and copper; these materials can be stretched into wire.

Brittle materials (e.g. glass, cast iron, ceramics) have virtually no ductility and fracture at the elastic limit.

Elastic materials include many plastics.

The main modes of failure that can arise from excessive stress on metal components are:

- Metal fatigue.
- Ductile failure.
- Brittle failure.
- Buckling.
- Corrosive failure.

Metal Fatigue

Metal-fatigue failures are by far the most common type where conditions producing mechanical vibration occur. Fluctuating-stress conditions produced by the vibration (e.g. an aircraft wing going up and down during flight) can cause the formation of a crack that propagates through the material. The crack reduces the area of material resisting the stresses until such time as the remaining material can no longer resist the stresses and fails. This failure is generally a brittle failure as it is the less ductile materials that are used to resist cyclic loading (as resistance to deformation is the key factor affecting this choice). The final failure is rapid and without warning. The rate of the cyclic loading on the material has little effect in relation to the chances of fatigue failure. Tests show that chances of fatigue occurring in a sample subject to 15,000Hz is about 10% greater than a similar sample exposed to 150Hz. A common feature of fatigue failure is that the initial crack always starts from the surface and penetrates into the body of the material. Surface blemishes, such as machining marks and foreign body inclusions, are all likely candidates to set a fatigue failure into action, as are holes for bolts, rivets, inspection hatches, etc.

Ductile Failure

Ductile fractures in metal usually occur when the yield stress of the material has been exceeded by the material being placed in tension. The metal, having moved into the plastic region of the stress/strain curve, loses its original shape. As the increase in length has to be accommodated by material from within the mass, there is a reduction in cross-sectional area which increases the true stress (it being inversely proportional to the area). This becomes the most highly stressed part and it is here that failure will occur. The failure will normally result from a single overload situation, i.e. a single event. There is no need for the system to have been under long-term overloads.

Brittle Failure

Brittle failures occur very suddenly and without warning, allowing a rapid release of energy, e.g. a load dropping from a crane. In simple terms, brittle fracture occurs because the structure of the material does not slip, either owing to the structure of the material itself or that insufficient time is available due to the intensity of the load. Small cracks spread through the material so quickly that a massive failure is produced. You should note that failure can occur when the applied stresses are below the yield stress.

The speed of failure often results in some of the energy in the material at failure being released as sound, giving the brittle failure a characteristic 'crack'. Note that in other failure modes, the actual failure may be by brittle mode (i.e. a parting of the material along the grain boundaries) but this will be only part of the sequence of failure (e.g. see ductile failure above).

Buckling

This is a term normally applied to the failure of a load-bearing part of a structure in compression. When a compressive force is put into a rod, bar or beam the force is resisted by the material. As the force increases, the material will tend to distort, preventing a straight transfer of the stress through the material. The next figure indicates how the shape of the material changes, and the effect of the stress.

An example of buckling can often be seen in steel-framed buildings that have been subject to fire. The heat of the fire leads to a weakening of the steel, which distorts under the compressive forces introduced by the loading of the roof, floors, etc. When cooled, it retains its distorted shape. Its ability to continue to be used to support a load would be dependent upon the amount of distortion and the effect of the fire on the integrity of the material itself.

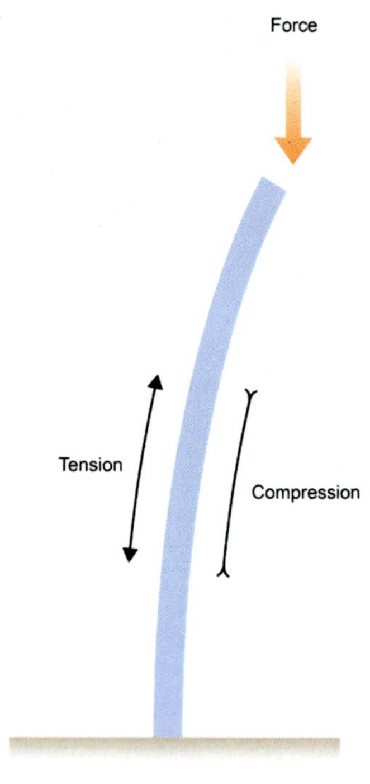

Rod under compression

Corrosive Failure

Corrosion is the effect of chemical change of the metal. Corrosion only affects metals as the process is one of ion exchange and occurs when positive and negative charges can move around. Corrosion occurs frequently in metals and is one of the most common forms of metal failure. It is difficult to predict due to the influence of a number of factors. The process involves the chemical conversion of the metal, which may occur in two main ways:

- **Oxidation** - occurs at high temperatures where the reaction of the metal with free oxygen is accelerated, which can rapidly 'eat away' the metal and weaken it. Due to the requirement of high temperatures (around 1,000°C) the situation is rare, but the speed and effect of it means it has to be taken into account in the design and selection of materials.
- **Electrochemical or galvanic** - occurs when the metal is subject to attack by 'free charged ions'. These come about by the presence of a solution, normally water, that acts as an 'electrolyte' between different metals or differently charged areas of the same material. The surface of a metal may become positively or negatively charged, which can become more pronounced depending on possible impurities, etc.

Advantages and Disadvantages of Non-Destructive Testing

Principles and Application

The aim of Non-Destructive Testing (NDT) is to test without having to destroy the integrity of the material or component.

NDT can easily be incorporated within a quality control regime during the manufacture of components with the advantage that substandard components are rejected at the earliest possible stage of production. Savings are therefore generated by not having work undertaken on components that will subsequently be rejected.

A number of NDT techniques are available, which are described below. Each has its limitations in use and it is important to understand the principles and limitations of each of the methods.

Non-destructive techniques involve leaving the test piece undamaged. Many of the techniques can be undertaken on site. The selection of the particular technique for a specific situation is important as it needs to take into account the limitations and weaknesses of each technique in order for there to be confidence in the results obtained.

Dye Penetrant

This technique involves the use of dye penetrant to highlight the defect so that a visual inspection can be made. The surface of the material to be tested is first cleaned thoroughly, including the removal of any lubricant in the form of oil or grease. In an on-site location, this is carried out using a solvent cleaner to remove any traces of grease. The dye is then applied, normally by spraying, and this penetrates into any surface crack. After a period of time the excess is wiped off. To make the dye visible, it is 'developed' by spraying with a developer which is naturally white but absorbs the dye from the defect, giving an indication on the surface as to the location of any defects at which to direct visual or other examinations.

Acoustic Emission

Acoustic Emission (AE) is the production of sound waves when a material undergoes internal change as a result of an external force. Elastic energy is released by a material as it undergoes deformation or fracture. Typical emission frequencies of these elastic waves are 100kHz to 1MHz. Mechanical loading on a component can generate elastic waves caused by small surface displacement of the material. The rapid release of energy within the material can be detected as cracks grow, fibres break or other damage occurs in the stressed material. Small-scale damage is detectable long before failure, so AE can be used as an NDT to find defects during structural proof tests and plant operation. The technique can also be used to study the formation of cracks during a welding process, as opposed to locating them after the weld has been formed with the more familiar ultrasonic testing technique.

Emissions are detected by an ultrasonic transducer probe in close contact with the material surface and transmitted to a receiving instrument which records and displays their frequency, amplitude and duration. If the characteristics of sounds normally emitted from the material are known then abnormal ones can be identified. If more than one probe is used, the location of the source of abnormal emission can be determined.

Typical applications of the AE principle in testing materials include:

- Examination of the behaviour of materials such as metals, ceramics, composites, rocks and concrete to examine crack propagation, fatigue, stress corrosion or creep.
- NDT during manufacturing processes.
- Continuous monitoring of metallic structures.
- Periodic testing of pressure vessels, pipelines, bridges and cables.

Ultrasonic

This method involves the use of a generator transmitting ultrasound waves into a material and detecting them when reflected from within the material. It is analogous to the geological mapping process that uses reflected shock waves to interpret discontinuity phenomena underground.

The ultrasound waves are generated in a head that is placed upon and moved across the surface, generally with some form of lubricant (e.g. water) to minimise any air gap. The ultrasound travels into the material and is 'bounced' back to a receiver mounted in the head. The output is read on an oscilloscope. The equipment must be calibrated to the full depth of the material before starting. Any defect will cause a variation in the return signal and this can be interpreted to indicate the depth of the defect, so it can detect defects within the material that do not show on the surface.

Radiography

This process requires a source of gamma or X-rays which are allowed to pass through the material and onto a strip of film. The radiation triggers a reaction in the film emulsion which, when developed (and known as 'radiographs'), shows where the material is sound and also where any discontinuity/defect exists within the material.

This is an important method where a permanent record of the inspection is required. It is used extensively in steel fabrication, particularly in highly specified situations such as welds on oil rigs, pipework and reactor vessels.

Eddy Current

When a high frequency AC current is passed through a coil, it sets up alternating magnetic fields. If the coil is placed next to the surface of an electrically conducting material it sets up eddy currents in the material. Any discontinuity in the surface causes a variation of the eddy current. If another coil is placed adjacent to the first, it can detect the changes in the eddy current, so indicating the location of a defect. These changes can be calibrated and used to determine the depth of any defect.

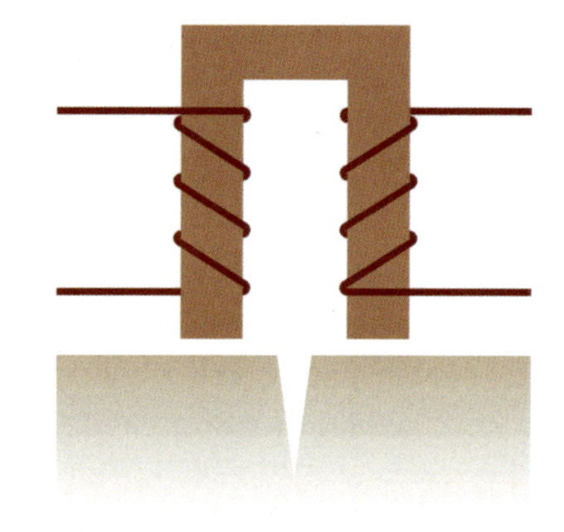

Eddy current testing

Magnetic Particle

This technique works by magnetising the component and applying magnetic particles or ink. Any defect in the component will show as it distorts the magnetic field and the particles lie 'differently'. The defect tends to cause a concentration of the magnetic field which attracts more particles to it than to the surrounding material.

The magnetic field may be applied using AC currents in one of a number of possible ways, as shown in the following figures:

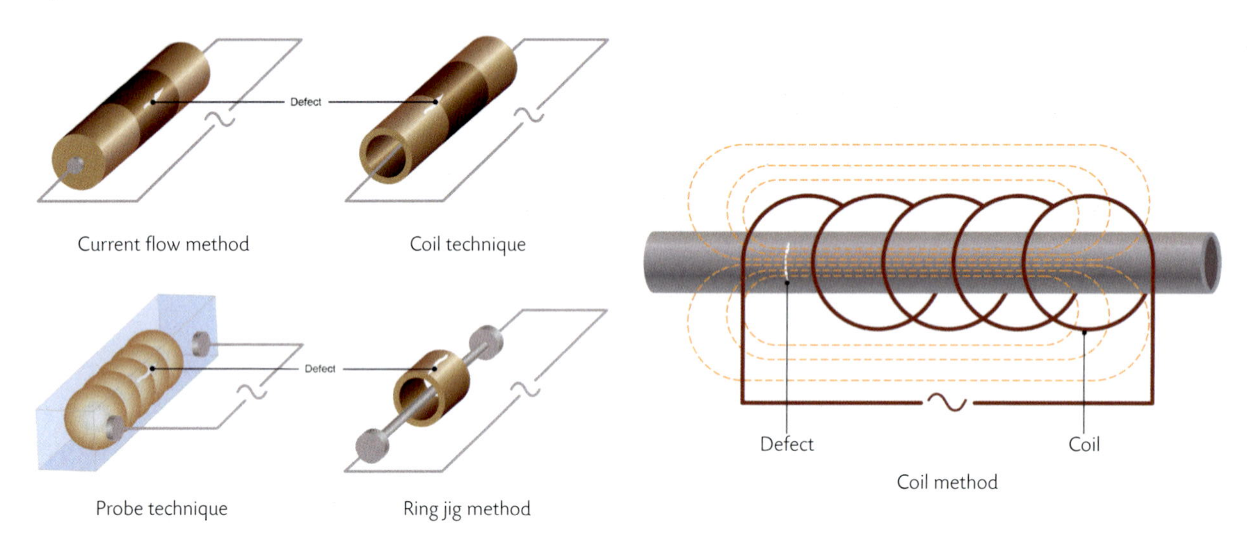

Circular magnetisation for longitudinal cracks

Magnetisation for transverse cracks

Summary of Advantages and Disadvantages

The following table summarises the NDT methods available, together with their advantages and disadvantages.

Summary of NDT methods

Test	Advantages	Disadvantages
Dye Penetrant	Cheap and convenient. Superior to visual examination. Can be used on all non-porous materials.	Surface defects only. Defects must be open to the surface.
Acoustic Emission	Can be used to study the formation of cracks during a process and precisely locate the source.	Relies on complex equipment and skilled operators.
Ultrasonic	Precise location of internal and external defects. Sizing of many defects possible.	Expensive equipment. Dependent on a skilled operator and a power supply.
Radiography	Permanent, pictorial, easily interpreted images obtained. Locates majority of internal defects.	Safety hazards. Expensive X-ray sets. Thickness limits (more so with X-rays). Power supply required.
Eddy Current	Rapid detection of surface or sub-surface flaws. Can measure depth of shallow flaws.	Cannot operate close to other free surfaces, e.g. thin sheet. Cannot find deep flaws. Requires power source.
Magnetic Particle	More sensitive than dye penetrant. Can also find sub-surface defects.	Ferrous metals only. Cannot find defects at any significant depth. Requires power source.

STUDY QUESTIONS

1. Outline the:

   (a) **advantages** and

   (b) **disadvantages**

   of the operation of a planned preventive maintenance system.

2. Outline the factors to be considered in developing a planned maintenance programme for safety-critical components.
3. Describe the factors to be considered when determining inspection regimes for work equipment.
4. Describe the principle of operation of ultrasound as a non-destructive testing method.
5. Choose two other forms of non-destructive testing (i.e. not including ultrasound). For each of the two NDTs chosen, outline the advantages and disadvantages.
6. If equipment has to be running or working during a maintenance operation and this presents risks, what measures can be taken to operate the equipment in a way that reduces the risk?
7. It may be necessary to isolate machinery from potential uncontrolled energy sources during repair, service or maintenance work.

   (a) Give examples of typical energy sources.

   (b) List possible isolation measures.

(Suggested Answers are at the end.)

Summary

Work Equipment and Machinery Maintenance

We have:

- Identified the hazards and control measures associated with maintenance of work equipment and machinery.
- Identified the statutory duties for maintenance under **PUWER**, **LOLER**, **COSHH**, **PSSR**, **Carriage of Dangerous Goods and Use of Transportable Pressure Equipment Regulations 2009**, **Personal Protective Equipment at Work Regulations 1992** and **Electricity at Work Regulations 1989**.
- Considered the three different strategies for maintenance, including the factors to consider, advantages and disadvantages of:
  - Planned preventive maintenance.
  - Condition-based maintenance.
  - Breakdown maintenance.
- Considered the factors to be considered in developing a planned maintenance programme including:
  - Importance in the process.
  - Machine complexity.
  - Relationship with other machines.
  - Availability of replacement equipment.
  - Identification of critical components.
  - Environmental factors.
  - Maintenance capability and design of the machine.
  - Maintainability within the design of the machine.
- Examined the factors to be considered in determining inspection regimes having consideration of the type of equipment, where it is used, how it is used and the method of inspection (including the use of new technologies such as remotely-operated vehicles and drones).
- Discussed the need for functional testing of safety related parts, including interlocks, protection devices, controls and emergency controls.
- The means by which machinery is safely set, cleaned and maintained including: safe systems of work, permits, isolation and procedures for working at unguarded machinery.
- Considered the means by which machines are isolated from all energy sources.
- Outlined the typical causes of material failure such as excessive stress, abnormal external loading, metal fatigue, ductile failure, brittle failure, buckling and corrosive failure.
- Considered the operation of the principal non-destructive testing techniques, which are dye penetrant, acoustic emission, ultrasonic, radiography (gamma and X-ray), eddy current and magnetic particle imaging.

Learning Outcome 11.7

NEBOSH National Diploma for Occupational Health and Safety Management Professionals

ASSESSMENT CRITERIA

- Understand why and how risks from working equipment and pressure systems should be managed.

LEARNING OBJECTIVES

Once you've studied this Learning Outcome, you should be able to:

- Explain how risks to health and safety arising from the use of work equipment are controlled.
- Outline the criteria for the selection of suitable work equipment for particular tasks and processes to eliminate or reduce risks.
- Outline the maintenance and prevention strategies when working with pressure systems.

Contents

Work Equipment

IN THIS SECTION...

- The selection of suitable work equipment for particular tasks and processes makes it possible to reduce or eliminate many risks to the health and safety of people in the workplace.
- Employers have a duty to ensure that work equipment is fit for purpose and is:
  - suitable;
  - safe for use;
  - maintained in a safe condition;
  - inspected; and
  - accompanied by suitable safety measures.
- Equipment must be suitable, by design, construction or adaptation, for the actual work it is provided to do.
- Energy or substances used or produced by equipment may present a risk to safety or health and should be supplied and/or removed in a safe manner.
- Well-designed layout of controls will reduce the ergonomic and anthropometric strains on the operator that could affect operator error type and frequency.
- One of the most effective ways of reducing the risk of harm from machinery is to eliminate, or reduce, the need for interaction between the operator and the machine.
- The size of openings, height of barriers and distance from danger are important considerations in safeguarding work equipment, as they affect whether a person can come into contact with dangerous parts of machinery.

Why Risk Assessments Must be Carried Out on Work Equipment

Apart from the legal argument for ensuring the health, safety and welfare of employees, an employer also owes a common law duty of care to his workforce and a moral duty to provide a workplace and work equipment that is safe.

Work equipment risk assessment

When the **Health and Safety at Work, etc. Act** was introduced in 1974, around 700 workers were dying at work every year and many thousands were suffering life changing non-fatal injuries including broken bones, amputations and burns. Many of these injuries were caused by poorly designed and unguarded work equipment and it wasn't until the advent of the early **Factory Acts** of 1833 (which was largely ignored by factory owners) and 1844, that all dangerous machinery was to be securely fenced off (guarded), and failure to do so would be regarded as a criminal offence.

HSE statistics, based on figures from the Labour Force Survey, have established there has been an estimated reduction of 90% in the number of fatal injuries since the introduction of the **1974 Act** with a corresponding reduction in the number of non-fatal injuries (although the figures do show a worrying flattening out of the numbers over the last 10 years).

PUWER sets out important health and safety requirements for the provision and safe use of work equipment which include:

- Management duties covering the selection of suitable equipment, maintenance, inspection, specific risks, information, instructions and training and the conformity of work equipment with legislation on product safety.
- Physical requirements for guarding of dangerous parts, the provision of appropriate stop and emergency stop controls, stability, lighting and suitable warning markings or devices.

Although risk assessment is not a stated requirement of **PUWER**, it is required by the **Management of Health and Safety at Work Regulations 1999 (MHSWR)** and implied in order to achieve compliance with **PUWER**.

Notwithstanding, the ACoP and guidance to **PUWER** recommends that risks to health and safety should be assessed taking into account matters such as the type of work equipment, substances and electrical or mechanical hazards to which workers may be exposed.

In broad terms, the employer must ensure that work equipment is:

- Suitable for the intended use.
- Safe for use, maintained in a safe condition and, in certain circumstances, inspected to ensure this remains the case.
- Used only by people who have received adequate information, instruction and training.
- Accompanied by suitable safety measures, e.g. protective devices, markings and warnings.

This may be achieved through risk assessment and the implementation of appropriate controls.

MORE...

HSE statistics for Great Britain can be found here:

www.hse.gov.uk/statistics/history/historical-picture.pdf

Employers' Duty to Ensure that all Work Equipment is Fit for Purpose

Regulation 4 of **PUWER** requires the employer to ensure that work equipment is constructed or adapted to be **suitable** for the purpose for which it is used or provided. The selection of suitable work equipment for particular tasks and processes makes it possible to reduce or eliminate many risks to the health and safety of people in the workplace. This applies both to the use and maintenance of the equipment.

Equipment needs to be suitable for the operation

When selecting work equipment, employers should also consider environmental conditions, such as:

- Lighting, temperature and humidity.
- Problems such as accelerated wear and exposure to wet or cold, which might be caused by using the equipment outside in poor weather conditions.
- Other work being carried out in the vicinity that may be causing noise, vibration or dust emissions, which may affect the operation.
- The activities of people who are not at work but who may be affected by noise, dust or fumes produced by the operation of the equipment.

Employers must assess the location where the work equipment is being used and take account of any risks that may arise from the particular circumstances.

The risks involved may mean that work equipment may not be able to be used in a particular place. For example, electrically-powered equipment is not suitable for use in wet or flammable atmospheres unless it is designed for this purpose. In such circumstances, employers should consider selecting suitably protected electrical equipment or alternative pneumatically- or hydraulically-powered equipment.

Means to Supply Energy or Substances

Another factor to consider when determining the suitability of work equipment is the possibility that energy used or produced by the equipment, such as electrical or mechanical energy, may present a risk to safety or health. Similarly, any substances used or produced by the equipment, such as fuels or process materials, may introduce similar risks.

The **Provision and Use of Work Equipment Regulations 1998** *Approved Code of Practice and Guidance* (L22) requires that:

> *"When determining the suitability of work equipment, you should ensure that where appropriate:*
>
> *(a) all forms of energy used or produced; and*
>
> *(b) all substances used or produced;*
>
> *can be supplied and/or removed safely."*

An example of this is a petrol engine generator discharging exhaust fumes into an enclosed space; where the employer has a duty to ensure the fumes are extracted from the area so employees are not put at risk of harm from the carbon monoxide.

MORE...

L22 - *Safe use of work equipment - Provision and Use of Work Equipment Regulations 1998 - Approved Code of Practice and Guidance* is available to download in full from:

www.hse.gov.uk/pubns/priced/l22.pdf

Ergonomic, Anthropometric and Human Reliability Considerations

DEFINITIONS

ERGONOMICS

The relationship between a person, the tools and equipment that they are using and the environment in which they are working.

Ergonomics is frequently used as a way of improving efficiency and in the context of occupational health and safety management is usually linked to the prevention of musculoskeletal disorders and improvements in reliability.

ANTHROPOMETRICS

The measurement of the human body.

In the context of occupational health and safety this is often related to the selection of appropriate equipment, such as desk and chairs. Understanding average dimensions within a population also informs decisions about guarding standards.

DEFINITION

HUMAN RELIABILITY

The degree to which people can be expected to perform to a specific standard.

Modern health and safety practice recognises that people are unable to perform to standard 100% of the time and human errors and violations will occur.

Layout and Operation of Controls and Emergency Controls

Before designing equipment or machinery, it is important to know what it is required to do and the control panel is no exception. Indeed, due to the effect that it can have on an individual's health, it is even more important to know how the individual will interface with the machine. The use of task analysis will assist in this process of identification.

For the man-machine (ergonomic) relationship to be effective, the human element needs to receive information relating to the process, performance and adverse conditions that require action (including emergency situations). In order to do this and so that the necessary action is taken, the design of the interface has to take into account the human characteristics (anthropometrics) that will allow the information to be received, processed and acted upon. The information can be received in a number of ways but generally it is either **visual** or **audible**.

There are several types of control available for any piece of machinery. What is important is that the chosen control is appropriate for the operation of the machine and that it does its job effectively and with minimum opportunity for error. The more important criteria that controls should meet, summarised from the Essential Health and Safety Requirements set out in Schedule 2 of the **Supply of Machinery (Safety) Regulations 2008**, are as follows:

- Clearly visible.
- Appropriately marked.
- Positioned for safe operation.
- Designed so that movement of the control is consistent with the effect required.
- Located outside the danger zone, except for certain controls, e.g. emergency stop.
- Positioned so that their operation cannot cause additional risk.
- Designed to prevent unintentional operation.
- Made to withstand foreseeable strain, particularly emergency controls.
- Starting of the machine can only be by means of the control, particularly after adverse conditions, e.g. a power cut.
- Stopping devices to be fitted.
- Emergency stopping devices to be fitted, clearly marked and to work effectively.
- The control must override any part of the system except the emergency control.
- Any fault or failure must not lead to danger.

MORE...

The **Supply of Machinery (Safety) Regulations 2008** are available to view in full at:

www.legislation.gov.uk/uksi/2008/1597/contents/made

There are various ways to prevent accidental operation of controls, such as:

- Recessing the control.
- Orientating the control so that the normal direction from which any accidental activation may occur will not cause it to be operated.
- Using two-handed controls and shrouded start buttons and pedals.
- Covering the control with a hinged cover.
- Locking the control.
- Operationally sequencing a set of controls.
- Increasing control distance.
- Ensuring that any starting that is initiated from a keyboard or other multifunction device, should require some form of confirmation in addition to the start command, and the results of the actuation should be displayed.

Controls have a number of characteristics that need to be taken into account. Remember that they are the prime man-machine interface and must be appropriate for the operator. For example, it is not appropriate to install a lever that requires a person to use his/her full body weight when the operator is required to be in the sitting position, making operation difficult, if not impossible. The following is a list of control characteristics:

- Displacement (either linear (up and down or side to side) or angular (rotational)).
- Operating force.
- Friction, inertia or drag.
- Number of positions.
- Direction of movement.
- Predetermined stops (detents).
- Appropriate identification.
- Compatibility with displays.
- Size.

Each of the above characteristics will, to a greater or lesser extent, affect the choice of the controls for any specific task.

Another important aspect for the selection of appropriate controls is their **location**. Key factors in this are:

- Number of controls - the fewer there are, the less chance of the wrong one being operated.
- Arranging controls to encourage a range of postures for the operator, which allow movement to keep the body 'fresh'.
- Arranging controls so that the sequence of operations is in an arc, so that the control layout is representative of the process.
- Where large forces are required to be exerted, use foot pedals or have power assistance.
- Have a clear distinction between normal and emergency controls.
- Keep consistent groupings of displays and controls; ideally, the display should be above the control.
- Prevent accidental operation by recessing or shielding the control or by distance to prevent them being knocked by an elbow, etc.

Selection of the appropriate control type is therefore important from the operational viewpoint.

The following table summarises the types of control and their characteristics:

Characteristics and uses of controls

Control Type	Use at Speed	Accuracy	Mounting Space Required	Use in an Array	Ease to Check Reading in Array
Toggle switch	Good	Good	Small	Good	Good
Rotary switch	Good	Good	Small	Good	Fair
Push button	Very good	Very poor	Small	Good	Poor
Rotary selector	Good	Good	Medium	Good	Good
Knob	Fair	Fair	Small-medium	Poor	Good
Handwheel	Poor	Good	Large	Poor	Poor
Crank	Fair	Poor	Medium-large	Poor	Poor
Lever	Good	Fair	Large	Good	Good
Foot pedal	Good	Poor	Large	Poor	Poor

Many systems do not just have one control and/or indicator, but a number of them on the same control panel. These will be arranged into an 'array', which introduces further problems. The effectiveness of controls in arrays is given in the table above.

Some key points relating to arrays are:

- Keep the control next to the indicator that gives the reading of its output so that the effect of operation of the control can be easily monitored.
- Have switches all set the same way, e.g. 'up' for off and 'down' for on (this reflects the stereotyping obtained by individuals from switching domestic lights on and off).
- Arrange for knobs all to increase by turning the same way.
- If a bank of dials is introduced, have them so that they all point the same way when normal. Any abnormal reading then stands out by position rather than the operator having to read all the dials.
- Where dials all have differing locations in the 'normal' position, then colour code the dial face to show when the needle is outside the norm. A glance will then highlight whether further attention to this particular information is required.
- Identify stereotypes for controls and take these into account. (Remember that other countries have different stereotypes and imported machines may be different, e.g. a light switch in the USA is 'on' when up rather than 'down' as in the UK.)
- Controls need to be located so that there is enough space between them that they can be grasped and operated.
- Well-designed layouts of controls will assist in the introduction of a man-machine interface that will not impose undue pressures and strains on the operator that could affect operator error type and frequency.

Reducing Need for Access

One of the most effective ways of reducing the risk of harm from machinery is to eliminate, or reduce, the need for interaction between the operator and the machine. Two successful methods used in industry for achieving this are:

- **Automation**

  In certain operations, usually routine, repetitive tasks, it is possible to replace workers with automatic systems such as robots, automatic guided vehicles, etc. These computer-controlled devices carry out the tasks without the associated human failings of fatigue, misjudgment or lapse of concentration. While an automated system requires maintenance of the machinery and the operating software, the day-to-day interaction is reduced to a minimum. Further control can be achieved by restricting access to automated areas, and the use of suitable safeguarding such as fencing, interlocking gate locks, pressure mats, etc.

Automation in the car industry

- **Remote Systems**

  This simply refers to the operation of a system or single piece of machinery by a person, or persons, who may not be in close proximity to that equipment. Practically everybody is familiar with the TV remote control which allows all functions to be carried out from the other side of the room. The same principle of 'remote control' is used in many different industrial settings from simple arrangements, such as a pendant control for an overhead crane, to more complex control rooms from which all routine operations are controlled and monitored. The obvious advantage is in segregating the operator from the immediate working area. However, total reliance on the controls could be seen as a disadvantage.

While it is clear that both of these systems can play a large part in reducing regular interaction (and therefore the risk of injury), maintenance, repairs and servicing still need to be carried out, and it is important to have rigorous procedures to cater for these activities.

Size of Openings, Height of Barriers and Distance from Danger

A person can come into contact with dangerous parts of machinery in two ways: either by reaching over the protective structure, or putting part of their body through openings in the protective structure.

Size of Openings

Arms or fingers can go through openings and legs and feet can be at risk from dangerous parts of machinery. BS EN ISO 13857:2019 *Safety of machinery - Safety distances to prevent hazard zones being reached by upper and lower limbs* superseded both BS EN 294 (the standard relating to upper limbs) and BS EN 811 (lower limbs) and contains tables and data to enable guards to be designed with an acceptable combination of height, horizontal distance from the hazard, and aperture size (for guards with mesh infill, or other openings in guards). The following tables illustrate some of these principles:

Clearance through openings

Limitation of Movement	Safety Distance (sr)	Illustration
		Dimensions in millimetres
Limitation of movement only at shoulder and armpit	≥ 850	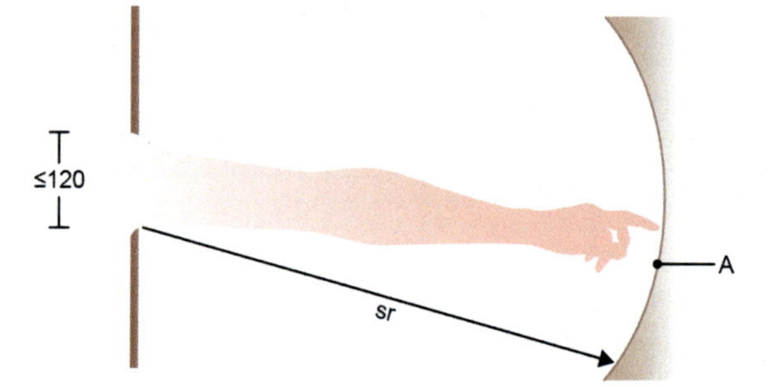
Arm supported up to elbow	≥ 550	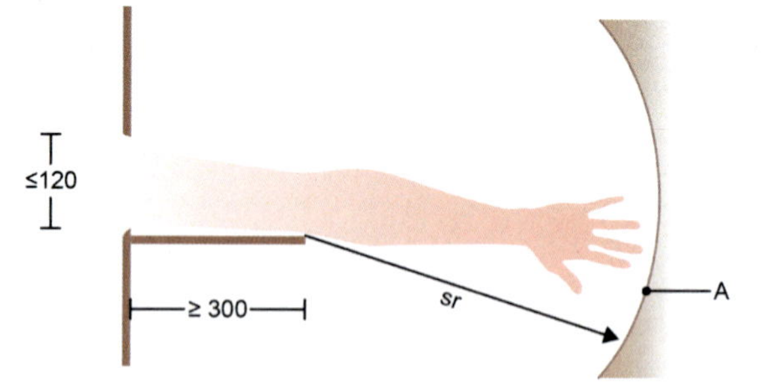
Arm supported up to wrist	≥ 230	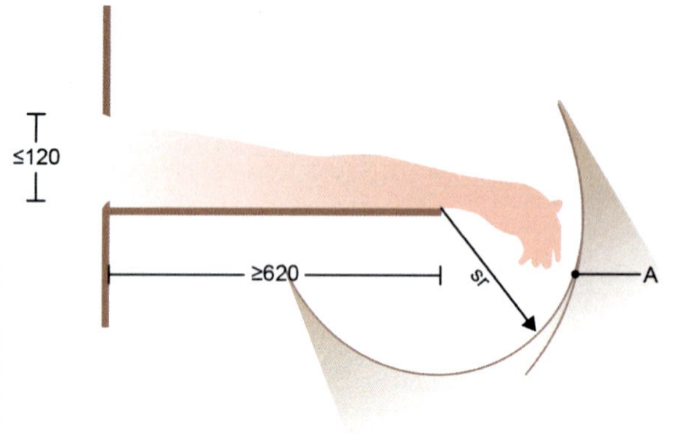
Arm and hand supported up to knuckle joint	≥ 130	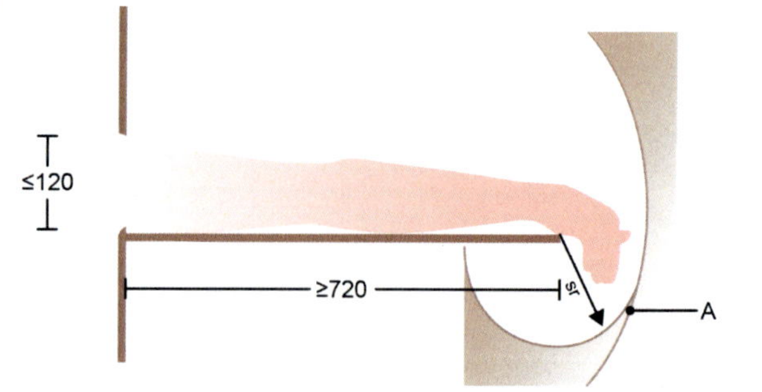

A: Range of movement of the arm

Source: Based on BS EN ISO 13857:2019 *Safety of machinery - Safety distances to prevent hazard zones being reached by upper and lower limbs*

Where **fingers** can go through the opening, then the clearances are as shown in the following table:

Clearance for fingers

Part of body	Illustration	Opening	Safety distance to hazard zone, s_r		
			Slot	Square	Round
Fingertip		$e \leq 4$	≥2	≥2	≥2
		$4 < e \leq 6$	≥10	≥5	≥5
Finger up to knuckle joint		$6 < e \leq 8$	≥20	≥15	≥5
		$8 < e \leq 10$	≥80	≥25	≥20
Hand		$10 < e \leq 12$	≥100	≥80	≥80
		$12 < e \leq 20$	≥120	≥120	≥120
		$20 < e \leq 30$	≥850[a]	≥120	≥120
Arm up to junction with shoulder		$30 < e \leq 40$	≥850	≥200	≥120
		$40 < e \leq 120$	≥850	≥850	≥850

NOTE The bold lines within the table delineate that part of the body restricted by the opening size.

[a] If the length of the slot opening is ≤65 mm, the thumb will act as a stop and the safety distance may be reduced to ≥200 mm.

Source: Based on BS EN ISO 13857:2019 *Safety of machinery - Safety distances to prevent hazard zones being reached by upper and lower limbs*

Where movement is restricted in some way by part of the protective structure, then dimensions may be reduced.

Height of Barriers

It may be that the danger is above or below the level of the person's shoulder, i.e. with the arm horizontal. As the arm would swing in an arc, the horizontal distance could be reduced, which is shown in the charts. The next figure shows the key dimensions referred to in the tables.

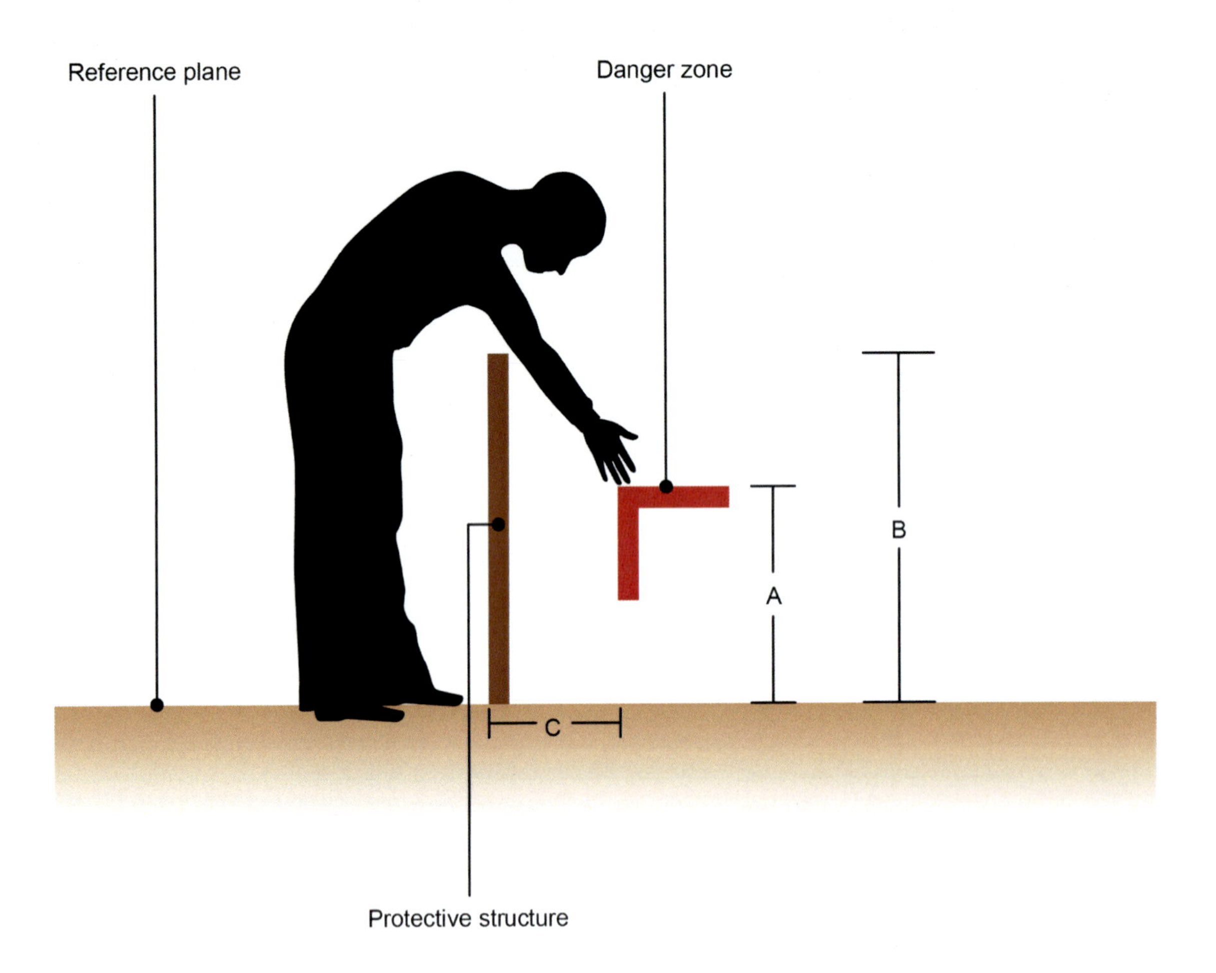

A Height of danger zone
B Height of protective structure
C Horizontal distance to danger zone

Key dimensions for the height of barriers

Source: Based on BS EN ISO 13857:2019 *Safety of machinery - Safety distances to prevent hazard zones being reached by upper and lower limbs*

Distance from Danger

The decision on minimum guarding distance is made on the basis of the risk from the danger that is to be avoided. Where the risk is low or high, the relevant distances are shown in the tables that follow.

Minimum distance for low risk

Dimensions in millimetres									
Height of Danger Zone	**Height of Protective Structure** [1]								
	1,000	1,200	1,400	1,600	1,800	2,000	2,200	2,400	2,500
	Horizontal Distance to Danger Zone								
2,500	-	-	-	-	-	-	-	-	-
2,400	100	100	100	100	100	100	100	100	-
2,200	600	600	500	500	400	350	250	-	-
2,000	1,100	900	700	600	500	350	-	-	-
1,800	1,100	1,000	900	900	600	-	-	-	-
1,600	1,300	1,000	900	900	500	-	-	-	-
1,400	1,300	1,000	900	800	100	-	-	-	-
1,200	1,400	1,000	900	500	-	-	-	-	-
1,000	1,400	1,000	900	300	-	-	-	-	-
800	1,300	900	600	-	-	-	-	-	-
600	1,200	500	-	-	-	-	-	-	-
400	1,200	300	-	-	-	-	-	-	-
200	1,100	200	-	-	-	-	-	-	-
0	1,100	200	-	-	-	-	-	-	-

[1] Protective structures less than 1,000mm high are not included because they do not sufficiently restrict movement of the body.

Source: Based on BS EN ISO 13857:2019 *Safety of machinery - Safety distances to prevent hazard zones being reached by upper and lower limbs*

Minimum distance for high risk

Dimensions in millimetres										
Height of Danger Zone	Height of Protective Structure [1]									
	1,000	1,200	1,400[2]	1,600	1,800	2,000	2,200	2,400	2,500	2,700
	Horizontal Distance to Danger Zone									
2,700	-	-	-	-	-	-	-	-	-	-
2,600	900	800	700	600	600	500	400	300	100	-
2,400	1,100	1,000	900	800	700	600	400	300	100	-
2,200	1,300	1,200	1,000	900	800	600	400	300	-	-
2,000	1,400	1,300	1,100	900	800	600	400	-	-	-
1,800	1,500	1,400	1,100	900	800	600	-	-	-	-
1,600	1,500	1,400	1,100	900	800	500	-	-	-	-
1,400	1,500	1,400	1,100	900	800	-	-	-	-	-
1,200	1,500	1,400	1,100	900	700	-	-	-	-	-
1,000	1,500	1,400	1,000	800	-	-	-	-	-	-
800	1,500	1,300	900	600	-	-	-	-	-	-
600	1,400	1,300	800	-	-	-	-	-	-	-
400	1,400	1,200	400	-	-	-	-	-	-	-
200	1,200	900	-	-	-	-	-	-	-	-
0	1,100	500	-	-	-	-	-	-	-	-

[1] Protective structures less than 1,000mm high are not included because they do not sufficiently restrict movement of the body.

[2] Protective structures lower than 1,400mm should not be used without additional safety measures.

Source: Based on BS EN ISO 13857:2019 *Safety of machinery - Safety distances to prevent hazard zones being reached by upper and lower limbs*

When the risk is overhead, the minimum dimension for low risk is 2,500mm. For high risk this is increased to 2,700mm. When this dimension is not possible, other protection should be provided, such as a fixed guard or mesh of such size as to prevent the individual reaching into the danger.

Risks Associated with Using Work Equipment

Because of the general risk assessment requirements in **MHSWR**, there is no specific regulation requiring a risk assessment in **PUWER**. However, the ACoP and guidance to **PUWER** recommends that risks to health and safety should be assessed taking into account matters such as the type of work equipment, substances and electrical or mechanical hazards to which people may be exposed.

TOPIC FOCUS

PUWER states in Regulation 4(2) that:

"In selecting work equipment, every employer shall have regard to the working conditions and to the risks to the health and safety of persons which exist in the premises or undertaking in which that work equipment is to be used ***and any additional risk*** *posed by the use of that work equipment".*

This is an implied duty (rather than express), further reinforced by Regulation 3 of **MHSWR**, to carry out machinery and work equipment risk assessments.

In order to comply with the requirement in **PUWER** to ensure the suitability of work equipment, it is necessary to consider the safety of work equipment from the following aspects:

- **Initial Integrity**

  Equipment must be suitable, by design, construction or adaptation, for the actual work it is provided for to do. When first providing work equipment for use in the workplace, the employer should ensure that it has been made to the requirements of the legislation, implementing any product Directive that is relevant to the equipment. This means that in addition to specifying that work equipment should comply with current health and safety legislation, it should also comply with the legislation implementing any relevant EU Directive. Where appropriate, the equipment should bear a CE marking and be accompanied by the relevant certificates or declarations as required by relevant product Directives. Adequate operating instructions should be provided with the equipment and adequate information about residual hazards, such as noise and vibration, should be provided. The equipment should also be checked for obvious faults.

- **Location Where It Will Be Used**

  The location in which the work equipment is to be used should be assessed to take account of any risks that may arise from the particular circumstances. Such factors can invalidate the use of work equipment in a particular place. For example, electrically-powered equipment is not suitable for use in wet or flammable atmospheres unless it is designed for this purpose. As mentioned previously, in such circumstances suitably protected electrical equipment or alternative pneumatically - or hydraulically - powered equipment would be a better choice.

  In addition, work equipment itself can sometimes cause risks to health and safety in particular locations which would otherwise be safe. Such an example is a petrol engine generator discharging exhaust fumes into an enclosed space.

- **Purpose for Which It Will Be Used**

  This requirement concerns each particular process for which the work equipment is to be used and the conditions under which it will be used. The equipment should be suitable for the process and conditions of use, for example:

  - A circular saw is generally not suitable for cutting a rebate, whereas a spindle moulding machine would be suitable because it can be guarded to a high standard.
  - Knives with unprotected blades are often used for cutting operations where scissors or other cutting tools could be used, reducing both the probability and severity of injury.

- **Decommissioning and End of Life**

  Decommissioning usually occurs at the end of the equipment life, when it is either no longer required, or is to be moved and perhaps re-commissioned at another location. Decommissioning larger pieces of production equipment or machinery may also have extra problems associated with the product itself, and its hazardous nature.

There is also a cost associated with equipment decommissioning because installation of the equipment in the first place is a much more straightforward process than taking it apart at the end of its working life and a detailed plan of action will be required. This plan should include:

- A risk assessment for all of the stages of disassembly taking into account isolation of the energy sources involved (electrical, hydraulic or pneumatic) and how any residual energy encountered during the work might be dealt with.
- Whether a specialist equipment or assistance will be required for lifting, cutting or containment of fluids or liquids found in the equipment that might leak during disassembly.
- If the work equipment is liable to leak or emit dangerous chemicals, flammable liquids or vapours then a safe system of work (permit to work) should be in place and all employees involved in the disassembly and disposal should be trained and competent.
- Any specialist PPE that might be required at any stage to deal with chemicals contained in the equipment.
- Areas where work equipment is being prepared for decommissioning or scrap should be segregated and all other personnel excluded.

Dismantling is also completed in stages, so some of the equipment may need to be dismantled before it can be decontaminated effectively. After decontamination, further dismantling may break the equipment down into smaller, more easily handled component parts for packaging and transportation. HSE guidance is included in **PUWER** and states:

> *"Work equipment should be erected, assembled or dismantled safely and without risk to health. Safe systems of work and safe working practice should be followed to achieve this. A safe system of work is a formal procedure which should be followed to ensure that work is carried out safely and is necessary where risks cannot be adequately controlled by other means.*
>
> *The work should be planned and hazards identified. You should ensure that the systems of work to be followed are properly implemented and monitored and that details have been communicated to those at risk."*

Every year an estimated 2 million tonnes of electrical and electronic items are discarded by householders and companies in the UK and this includes most products that have a plug or need a battery and fall into the category of work equipment. The **Waste Electrical and Electronic Equipment (WEEE) Regulations 2013** became law in the UK on the 1 January 2014 and although there are ten broad categories of WEEE outlined within the Regulations; for the purposes of this Unit, electrical and electronic tools in particular, such as drills, saws, sewing machines and electric lawnmowers used in an industrial capacity, are included.

Although decommissioning and end-of-life scrap tasks are often outsourced to an external contractor there are many, well documented instances of this practice going spectacularly wrong.

Significant Risk

DEFINITION

SIGNIFICANT RISK

A significant risk is one which could foreseeably result in a major injury or worse arising:

- from incorrect installation or re-installation;
- from deterioration; or
- as a result of exceptional circumstances which could affect the safe operation of the work equipment.

Inspection (as required by Regulation 6 of **PUWER**) is only necessary where there is a significant risk.

The ACoP and guidance to **PUWER** refers to circumstances whereby a risk assessment carried out under **MHSWR** identifies a significant risk to the operator or other workers from the installation or use of the work equipment, which would require a suitable inspection to be carried out.

Consequently, the risk assessment process for work equipment will need to consider not only the initial integrity of the equipment but also any risks arising from its initial installation and any subsequent re-installation. So, equipment that needs to be repeatedly installed then dismantled, for example, will need this aspect considered in the risk assessment.

Another consideration is deterioration of the equipment during use, either from defects, damage or wear, or possibly from the nature of the environment in which it is used, which could result in an unacceptable risk.

If the risk assessment is suitable and sufficient, as required by **MHSWR**, it should identify any exceptional circumstances which could affect the safe operation of the work equipment and therefore, although possibly of low probability, will still need to be considered in the risk assessment.

Risk Control Hierarchy

Risk control measures for work equipment follow the same principle as for all risk control strategies and are set out in the Topic Focus box below.

TOPIC FOCUS

The **hierarchy of controls** for work equipment:

- **Remove all risk by design**, i.e. introduce intrinsic safety.
- Use **fixed enclosed guards**; fixed guards have no moving parts which can fail or be abused.
- Use **other guards**, including movable guards, adjustable guards, automatic guards and fixed guards which are not fully enclosing.
- Use **protection devices** which do not prevent access but do prevent motion of the work equipment when close, e.g. pressure mats, infrared beams.
- Use **protection appliances** which hold or manipulate the workpiece, keeping the operator away from danger, e.g. a push stick on a circular saw.
- Provide **information**, **instruction**, **training and supervision**, which is always important and a requirement no matter what guarding arrangements are in place. It is particularly important when the risk cannot be adequately eliminated by the above, e.g. a hand-held electric drill.
- Use **PPE**.

MORE...

Further guidance on **PUWER** can be found in INDG291 *Providing and using work equipment safely – A brief guide*, available at:

www.hse.gov.uk/pubns/indg291.pdf

and also INDG229 *Using work equipment safely*, available at:

www.hse.gov.uk/pubns/indg229.pdf

These control measures can be simplified into the following strategy, which is found in the HSE guidance INDG291 on **PUWER**:

- Eliminate the risks created by the use of work equipment by careful selection.
- Employ 'hardware' (physical) measures such as:
  - Suitable guards.
  - Protection devices.
  - Markings and warning devices.
  - System control devices, e.g. emergency stop buttons.
  - PPE.
- Employ 'software' measures such as:
  - Following safe systems of work.
  - Ensuring maintenance is only performed when equipment is shut down.
  - Provision of information, instruction and training.

PPE used in conjunction with work equipment

INDG291 goes on to explain:

> *"A combination of these measures may be necessary depending on the requirements of the work, your assessment of the risks involved, and the practicability of such measures."*

Training and Competence

Competency can be defined as the ability to undertake responsibilities and perform activities to a recognised standard on a regular basis. It is a combination of skills, experience and knowledge.

Training is an important component of establishing competency but is not sufficient on its own. For example, consolidation of knowledge and skills through training is a key part of developing competency.

MORE...

Information on human factors, including training and competence, is available at:

www.hse.gov.uk/humanfactors

You will also find various links there to further sources of useful information on this topic.

Circumstances when Training is Likely to be Required

It is an important requirement of **PUWER** that all persons who use work equipment receive adequate health and safety training in how to use work equipment, the risks associated with such use, and the precautions that need to be taken.

The requirements for adequate training will vary according to the job or activity and the work equipment, but in general it will be necessary to:

- Evaluate the existing competence of employees to operate the work equipment in use.

- Evaluate the competence employees will need to manage or supervise the use of work equipment.
- Train the employee to make up any shortfall between their existing competence and that required.

Circumstances when training is likely to be required include at induction, where there are changes in work activities, where new equipment or technology is introduced, where the system of work changes, and when refresher training is necessary:

- **Induction**

  Training needs are likely to be greatest on recruitment. Recruitment and placement procedures should ensure that employees have the necessary abilities to do their jobs safely or can acquire them through training. New recruits need basic induction training into how to work safely as well as arrangements for first aid, fire and evacuation.

- **Changes in Work Activities**

  Additional training is required if the risks to which people are exposed change due to a change in their work activities. Consequently, people changing jobs or taking on extra responsibilities need to know about any new health and safety implications.

- **Introduction of New Technology or New Equipment**

  The risks to which people are exposed may change due to the introduction of new technology or new equipment. Again, persons employing new processes or new work equipment also need to know about the health and safety implications of such changes.

- **Changes in Systems of Work**

  Even if work activities and work equipment remain unchanged, it is likely that monitoring of health and safety standards and revision of risk assessments will result in improvements to systems of work which will need to be implemented through revised training of the relevant workforce.

- **Refresher Training**

  Refresher training should be provided if necessary, because skills decline if they are not used regularly. A key example of this is people who deputise for others on occasions, who will probably need more frequent refresher training than those who do the work regularly.

Groups of People Having Specific Training Needs

Supervisors

In addition to the requirement to train users of work equipment, the employer also has a responsibility to ensure that any employees who supervise or manage the use of work equipment receive adequate health and safety training in how to use the work equipment, the risks associated with such use, and the precautions that need to be taken.

Young and Vulnerable Persons

Training and proper supervision of young people is particularly important because of their relative immaturity and unfamiliarity with the working environment. Induction training is of particular importance since all employees should be competent to use work equipment safely regardless of their age.

MHSWR contain specific requirements relating to the employment of young people (under the age of 18), requiring employers to assess risks to young people before they start work, taking into account their inexperience, lack of awareness of potential risks and their immaturity.

Competence, External and Self-Supervision

The relationship between competence, external (imposed) supervision and self-supervision is illustrated in the following diagram taken from the previous version of HSG65 *Successful health and safety management*:

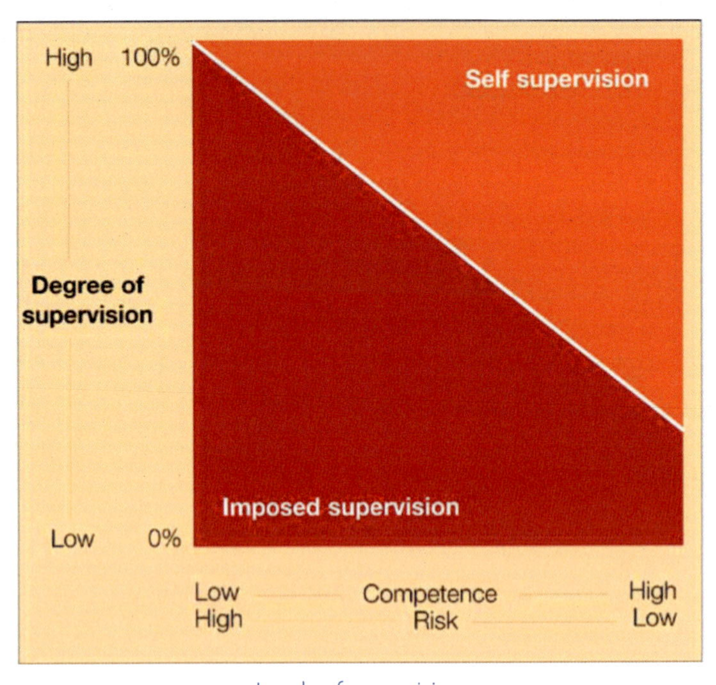

Levels of supervision
Source: HSG65 *Successful health and safety management* (2nd ed.), HSE, 1997 (now superseded)

The current version of HSG65 (third edition, 2013) adopts a Plan, Do, Check, Act approach and makes reference to extensive HSE guidance on competence.

MORE...

The HSE guidance on competence can be found at:

www.hse.gov.uk/managing/competence.htm

Further information can be found in HSG65 *Managing for health and safety*, available at:

www.hse.gov.uk/pubns/priced/hsg65.pdf

In deciding on the appropriate level of supervision for particular tasks, the level will depend on the risks involved as well as the competence of employees to identify and handle those risks. Consequently, external supervision will be needed if employees are new to a job, undergoing training or doing jobs which present special risks. Some supervision of fully competent individuals will always be needed to ensure that standards are being met consistently.

Circumstances Where There are Specific Training Needs for Certain Hazardous Types of Work Equipment

We have already noted that it is an important requirement of **PUWER** that all persons who use work equipment receive adequate health and safety training in how to use work equipment, the risks associated with such use, and the precautions that need to be taken. For certain hazardous types of work equipment there are specific training needs outlined below.

Self-Propelled Work Equipment

As with the training required for all work equipment, the training standard required for operators of self-propelled work equipment should be adequate in ensuring the health and safety of other workers and anyone else who may be affected by the work.

Self-propelled work equipment

Source: HSE WAIT Tool (www.hse.gov.uk)

The ACoP and guidance to **PUWER** (L22) specifically imposes minimum training obligations in relation to driver training and states:

> *"You should ensure that self-propelled work equipment, including any attachments or towed equipment, is only driven by workers who have received appropriate training in the safe driving of such work equipment."*

There is a further ACoP and guidance for those using lift trucks (L117). This supports the **PUWER** ACoP in dealing specifically with the training for rider-operated lift trucks and states that:

> *"Employers should not allow anyone to operate, even on a very occasional basis, lift trucks... who has not satisfactorily completed basic training and testing as described in this ACoP, except for those undergoing such training under adequate supervision."*

L117 also requires those providing the training to have undergone appropriate training in instructional techniques and skills assessment, and to have sufficient industrial experience and knowledge of working environments to put their instruction in context.

It should be noted that the duty to provide training under Regulation 9 of **PUWER** reinforces the general obligation under Section 2 of **HSWA** and the additional requirements on capabilities and training under Regulation 13 of **MHSWR**.

Training should take place during working hours and be at no cost to the employee. If it is necessary for training to take place outside the employee's normal working hours, this should be treated as an extension of their time at work.

Chainsaws

Chainsaws are potentially dangerous machines which can cause major injury if used by untrained people. Anyone who uses a chainsaw at work should have received adequate training and be competent in using a chainsaw for that type of work. The training should include:

Chainsaw

- Dangers arising from the chainsaw itself.
- Dangers arising from the task for which the chainsaw is to be used.
- The precautions to control these dangers, including relevant legal requirements.

The ACoP for **PUWER** sets a minimum standard for competence of people using chainsaws in tree work, requiring appropriate training and a relevant certificate of competence.

Woodworking Machines

The risks associated with the use of woodworking machinery are high since it relies on high-speed sharp cutters to do the job which, in many cases, are exposed to enable the machining process to take place. Additionally, many machines are still hand-fed.

Machine operators, those who assist in the machining process, and those who set, clean, or maintain woodworking machinery should be provided with training.

All training schemes should include the following elements:

- **General** - instruction in the safety skills and knowledge common to woodworking processes.
- **Machine specific** - practical instruction in the safe operation of the machine, including in particular:
  - The **dangers** arising from the machine and any limitations as to its use.
  - The main **causes of accidents** and relevant **safe working practices** including the correct use of guards, protection devices, appliances and the use of the manual brake where fitted.
- **Familiarisation** - on-the-job training under close supervision.

Power Presses

Power presses are among the most dangerous machines used in industry. Amputation or serious injury can result from accidents caused by trapping between the tools of a power press and the guarding mechanisms are subject to continuous wear.

Persons appointed to inspect power presses require training which includes suitable and sufficient practical instruction in relation to each type of power press and guard and/or protection device used.

Press operators are most likely to need training when they are recruited. However, training is also required if:

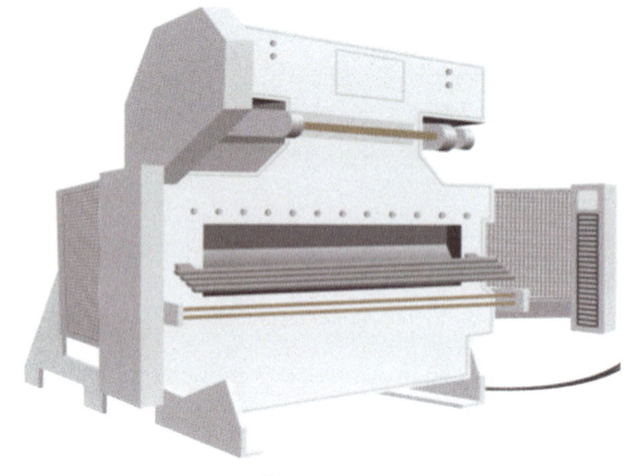

Power press

- The risks to which people are exposed change.
- New equipment or technology is introduced.
- The system of work changes.

Training is also required for people who supervise or manage the use of power presses. Such training should include the safe operation of the press and the risks posed to the person carrying out the work (e.g. the press operator, setter or appointed person) as well as the quality of the inspection and test carried out by the appointed person.

Abrasive Wheels

One of the main risks associated with the use of abrasive wheels is injury resulting from breakage. Accident statistics indicate that nearly half of all accidents involving abrasive wheels are due to an unsafe system of work or operator error. Consequently, training is required both in the use and in the mounting of abrasive wheels.

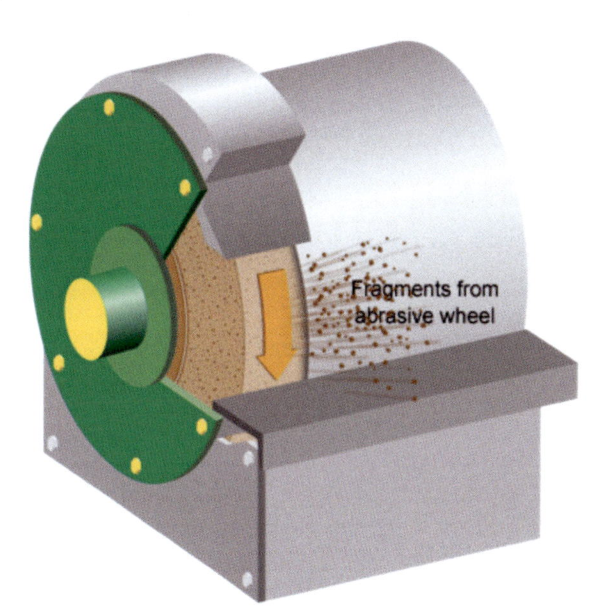

Abrasive wheel

Any training programme should cover at least the following:

- Hazards and risks arising from the use of abrasive wheels and the precautions to be observed.
- Methods of marking abrasive wheels with their type, size and maximum operating speed.
- How to store, handle and transport abrasive wheels.
- How to inspect and test abrasive wheels for damage.
- The functions of all the components used with abrasive wheels such as flanges, blotters, bushes, nuts.
- How to assemble abrasive wheels correctly to make sure they are properly balanced and fit to use.
- The proper method of dressing an abrasive wheel.
- The correct adjustment of the work rest on pedestal or bench grinding machines.
- The use of suitable PPE, e.g. eye protection.

Information Required for the Safe Use and Operation of Work Equipment

The requirement for providing information for the safe use and operation of work equipment is contained in Regulation 8 of **PUWER**.

Providing information for the safe use and operation of machinery

The employer has to ensure that all persons who use work equipment, and also those who supervise or manage it, have adequate health and safety information, or if necessary, written instructions on the use of the work equipment.

This information should cover:

- All health and safety aspects arising from the use of the work equipment.
- Any limitations on these uses.
- Any foreseeable difficulties that could arise.
- The methods to deal with them.
- Any additional information obtained from experience of using the work equipment.

Consequently, the employer has to make available all relevant health and safety information and, where appropriate, written instructions on the safe use and operation of machinery to their workforce. Workers should have easy access to such information and instructions and be able to understand them.

Such written instructions can include:

- Information provided by manufacturers or suppliers of work equipment such as instruction sheets or manuals, instruction placards, warning labels and training manuals.
- In-house instructions.
- Instructions from training courses.

Easily Understandable Information and Instructions

Regulation 8 of **PUWER** also requires the information and instructions to be easily understood by those concerned. Written instructions should be available to the people directly using the work equipment and also other appropriate people such as maintenance staff.

Supervisors and managers also need access to the information and written instructions. The amount of detailed health and safety information they will need to have immediately available for day-to-day running of production lines will vary, but it is important that they know what information is available and where it can be found.

Information can be verbal where this is considered sufficient, but where there are complicated or unusual circumstances, the information should be in writing. Other factors that need to be taken into consideration include:

- the degree of skill of the workers involved;
- their experience and training;
- the degree of supervision; and
- the complexity and length of the particular job.

The information and written instructions should:

- Be easy to understand.
- Be in clear English and/or other languages if appropriate for the people using them.
- Be set out in logical order with illustrations where appropriate.
- Use standard symbols where appropriate.

Special arrangements may be needed for employees with language difficulties or with disabilities which could make it difficult for them to receive or understand the information or instructions.

TOPIC FOCUS

Warnings

Regulation 24 of **PUWER** is concerned with **warnings** and requires the employer to ensure work equipment incorporates any warnings or warning devices needed for health and safety purposes, and that these warnings are unambiguous, easily perceived and easily understood.

Examples include:

- Notices such as:
  - Positive instructions (e.g. 'hard hats must be worn').
  - Prohibitions (e.g. 'not to be operated by people under 18 years').
  - Restrictions (e.g. 'do not heat above 60°C').
- Warning devices which are:
  - Audible (e.g. reversing alarms on construction vehicles).
  - Visible (e.g. a light on a control panel indicating that a fan on a microbiological cabinet has broken down or a blockage has occurred on a particular machine).
  - An indication of imminent danger (e.g. a machine about to start) or development of a fault condition (e.g. pump failure or conveyor blockage indicator on a control panel).
  - The continued presence of a potential hazard (e.g. hotplate or laser on).

STUDY QUESTIONS

1. Outline the factors that need to be considered when identifying the risks associated with the use of work equipment.
2. What is meant by the term 'significant risk' in relation to work equipment?
3. Describe the three elements of the risk control strategy relating to work equipment.

(Suggested Answers are at the end.)

Pressure Systems

IN THIS SECTION...

- The **Pressure Systems Safety Regulations 2000 (PSSR)** provide the definition of a pressure system.
- **PSSR** requires a written scheme of examination for the system.
- A strategy to prevent the failure of pressure systems should include: design and construction; repair and modification; information and marking; safe operating limits; written scheme of examination; maintenance and record keeping; and the requirement for competent persons.

Definition of a Pressure System

Regulation 2 of the **Pressure Systems Safety Regulations 2000 (PSSR)** describes a pressure system as:

" (a) a system comprising one or more pressure vessels of rigid construction, any associated pipework and protective devices;

(b) the pipework with its protective devices to which a transportable pressure receptacle is, or is intended to be, connected; or

(c) a pipeline and its protective devices, which contains or is liable to contain a relevant fluid, but does not include a transportable pressure receptacle".

Pressure system

DEFINITION

RELEVANT FLUID

This means:

- steam; or
- any fluid or mixture of fluids which is at a pressure greater than 0.5 bar above atmospheric pressure, and which is:
  - a gas; or
  - a liquid which would have a vapour pressure greater than 0.5 bar above atmospheric pressure when in equilibrium with its vapour at either the actual temperature of the liquid or 17.5°C; or
- a gas dissolved under pressure in a solvent contained in a porous substance at ambient temperature and which could be released from the solvent without the application of heat.

Examples of pressure systems and equipment include:

- Boilers and steam heating systems.
- Pressurised process plant and piping.

- Compressed air systems (fixed and portable).
- Pressure cookers, autoclaves and retorts.
- Heat exchangers and refrigeration plant.
- Valves, steam traps and filters.
- Pipework and hoses.
- Pressure gauges and level indicators.

Pressure systems can be divided into three categories:

- Minor systems include those containing steam, pressurised hot water, compressed air, inert gases or fluorocarbon refrigerants, which are small and present few engineering problems. The pressure (above atmospheric pressure) is less than 20 bar and the temperatures in the system should be between −20°C and 250°C.
- Intermediate systems include most storage and process systems which do not fall into either of the other two categories.
- Major systems are those which, because of their size, complexity or hazardous contents, require the highest level of expertise in determining their condition. They include steam-generating systems where the individual capacities of the steam-generators are more than 10MW, large pressure vessels and chemical reaction vessels.

Types of Inspection, Frequencies and Statutory Basis for Examination of Pressure Systems

Types of Inspection and Frequencies

Guidance contained in L122 *Safety of pressure systems - Pressure Systems Safety Regulations 2000* puts into context examples of the expected type and frequency of inspection carried out under the written scheme of examination and it is for the competent person appointed by the pressure system owner/operator to decide when and what type of inspection is required when the written scheme of examination is produced.

When deciding on the periodicity between examinations, the aim should be to ensure that sufficient examinations are carried out to identify at an early stage any deterioration or malfunction which is likely to affect the safe operation of the system. Different parts of the system may be examined at different intervals, depending on the degree of risk associated with each part.

There are many factors to be considered by the competent person when deciding the maximum interval between examinations under the written scheme of examination, and the competent person will use their judgment and experience to determine the appropriate interval based on the relevant information.

Earlier legislation such as the **Factories Act of 1961** is often used as a general rule by competent persons to decide maximum examination intervals. For steam boilers, the period was usually 14 months with more frequent examinations specified where operating conditions were arduous.

The examination period for steam receivers linked to such plant was generally in the range of 26-38 months. Air receivers on compressed air systems were generally examined every 24-48 months, with examinations taking place less frequently, i.e. every 72 months where corrosion was minimal and maintenance of safety standards was high.

Statutory Basis for Examination of Pressure Systems

Undertake the statutory inspection of the full system

The **Pressure Systems Safety Regulations 2000** cover pressure vessels, pipework and protective devices and place obligations on anyone who manufactures or constructs a new pressure system, and anyone who repairs or modifies a new or existing pressure system or part of it, to ensure that no danger will arise when it is operated within the safe operating limits specified for that plant.

The other main requirements of the Regulations are that the user must:

- Establish the safe operating limits of the system.
- Have a written scheme of examination for the system.
- Maintain the system.
- Have operating instructions and ensure that the system is only operated in accordance with the instructions.

The written scheme of examination (Regulations 8 to 10, 14) is important. It must be compiled by a competent person before a pressure system can be operated. Details of the pressure vessels, protection devices and pipework must be included in the scheme. It must specify the nature and frequency of examinations (this should be judgment-based - there are no specific requirements) and the measures necessary to prepare the system for safe examination.

A report of the periodic examination by the competent person must be given to the user or owner of the system within 28 days. However, if there is imminent danger from the continued operation of the system, the report must be provided within 14 days and a copy should be provided to the enforcing authority. The Regulations also require that records are retained.

Prevention and Testing Strategies

We know that if a piece of pressure equipment fails and bursts violently apart, the results can be devastating to people in its vicinity because parts of the equipment could be propelled at very high velocities over great distances, causing injury and damage to people and buildings hundreds of metres away.

TOPIC FOCUS

To **prevent** catastrophic failures like this, employers need to have a strategy in place to reduce the risk of this occurring. They need to know the:

- Pressure in the system.
- Type of liquid or gas and its properties.
- Suitability of the equipment and pipework that contains it.
- Age and condition of the equipment.
- Complexity and control of its operation.
- Prevailing conditions (e.g. a process carried out at high temperature).
- Skills, knowledge and experience of the people who maintain, test and operate the pressure equipment and systems.

To **reduce the risk** of a pressure system failure, employers should put in place precautions, such as ensuring:

- The system can be operated safely, for example without having to climb or struggle through gaps in pipework or structures.
- That after any repair or modification to the pressure system it is examined by a competent person before being allowed to come back into use.
- There is a set of operating instructions for all of the equipment in the system and for the control of the system as a whole, including in emergencies.
- There is a maintenance programme for the system which takes into account the system and equipment age, its uses and the environment in which it is being used.

How?

- Before using pressure equipment, employers should ensure they have a written scheme of examination if one is required. They also need to make sure that any inspections required have been completed by a competent person, and that the results have been recorded.
- Employers/operators should always operate the equipment within the safe operating limits. If these are not provided by the manufacturer or supplier, advice from a competent person, for example your employers' liability insurer, can be sought.
- Employers must provide instruction and relevant training for the workers who are going to operate the pressure equipment and also include what to do in an emergency.
- There must be an effective maintenance plan in place, which is carried out by appropriately trained people and any modifications are planned, recorded and do not lead to danger.

(Source: HSE.)

The key points in a prevention strategy are:

- Fit legally-required safety devices.
- Operate the system within the limits specified.
- Undertake the statutory inspection of the full system.
- Undertake more regular inspections between statutory inspections.
- Use non-destructive testing techniques to determine what is happening inside the system.

MORE...

More information can be found at:

www.hse.gov.uk/toolbox/pressure.htm

Design and Construction

Designers and manufacturers must consider at the manufacturing stage both the purpose of the plant and the means of ensuring compliance with the Regulations.

The designer, manufacturer, importer and supplier should consider and take due account of the following, where applicable:

- Expected working life (the design life) of the system.
- Properties of the contained fluid.
- All extreme operating conditions including start-up, shutdown and reasonably foreseeable fault or emergency conditions.
- The need for system examination to ensure continued integrity throughout its design life.
- Any foreseeable changes to the design conditions.
- Conditions for standby operation.
- Protection against system failure, using suitable measuring, control and protective devices as appropriate.
- Suitable materials used for each component part.
- External forces expected to be exerted on the system, including thermal loads and wind loading.
- Safe access for operation, maintenance and examination, including the fitting of access (e.g. door) safety devices or suitable guards, as appropriate.

Repair and Modification

When designing any modifications (including extensions or additions) or planning repairs to the pressurised parts of the system, whether temporary or permanent, the following should be taken into account:

- The original design specification.
- The duty for which the system is to be used after the repair or modification, including any change in relevant fluid.
- The effects any such work may have on the integrity of the pressure system.
- Whether the protective devices are still adequate.
- Continued suitability of the written scheme of examination.

Information and Marking

The aim here is to ensure that adequate information about any pressure system subject to **PSSR** is made available to users/owners by designers, suppliers or those who modify or repair equipment. Basic information about pressure vessels should be permanently marked on the vessel. The information required is:

- The manufacturer's name.
- A serial number.

- Date of manufacture.
- The standard to which the vessel was built.
- Maximum and minimum allowable pressure.
- Design temperature.

Additional information about pressure vessels and information relevant to the whole system should be provided in writing.

It is not possible to give a complete list of all the additional information which might be necessary, but the following items should be considered where relevant:

- Design standards used and evidence of compliance with relevant standards or documentation showing conformity.
- Design pressures (maximum and minimum).
- Fatigue life.
- Design temperatures (maximum and minimum).
- Creep life.
- Intended contents, especially where the design has been carried out for a specific process.
- Flow rates and discharge capacities.
- Corrosion allowances.
- Wall thickness.
- Volume capacities, especially for storage vessels. Depending on the intended contents, these may be expressed as maximum volume, pressure or filling ratio.
- Materials of construction.

Safe Operating Limits

The Regulations make the designer, manufacturer and supplier responsible for providing adequate information about the system or its component parts. They prohibit the user/owner from operating the system or allowing it to be operated before the safe operating limits have been established.

The exact nature and type of safe operating limits which need to be specified will depend on the complexity and operating conditions of the particular system. Small, simple systems may need little more than the establishment of the maximum pressure for safe operation. Complex, larger systems are likely to need a wide range of conditions specified, such as the maximum and minimum temperatures and pressures, nature, volumes and flow rates of contents, operating times, heat input or coolant flow. In all cases, the safe operating limits should incorporate a suitable margin of safety.

Written Scheme of Examination, Maintenance and Record Keeping

It is the responsibility of the user/owner to select a competent person capable of carrying out the duties in a proper manner with sufficient expertise in the particular type of system. In some cases, the necessary expertise will lie within the user/owner's own organisation.

The term 'competent person' is used in connection with two distinct functions:

- Drawing up or certifying schemes of examination.
- Carrying out examinations under the scheme.

Before a pressure system is operated, the user/owner must ensure that a **written scheme of examination** has been prepared. It should be drawn up by a competent person, or be drawn up by someone other than a competent person, who is certified as suitable by a competent person.

Responsibility under the Regulations may be summarised as follows:

- The user/owner ensures the scope of the scheme is appropriate, i.e. which parts of the system are covered (with advice, if necessary, from a suitably experienced adviser).
- The competent person specifies the nature and frequency of examinations and any special measures necessary to prepare the system for safe examination.

The purpose of **maintenance** is to ensure the safe operation and condition of the system. The actual process of carrying out the maintenance tasks is not covered. The risks associated with maintenance must be assessed to comply with the requirements of the **Management of Health and Safety at Work Regulations 1999** and the appropriate precautions.

The need for maintenance should not be confused with the requirement for examinations under the written scheme. They are two separate issues.

The type and frequency of maintenance for the system should be assessed and a suitable maintenance programme planned.

A suitable system for **recording and retaining information** about safe operating limits and any changes to them should be used. Whatever method is used, the information should be readily available to those people who need it, including the competent person responsible for the examinations in accordance with the written scheme.

For mobile systems, the owner must provide the user with a written statement detailing the safe operating limits or ensure that this information is clearly marked on the equipment.

Competent Persons

It is the responsibility of the user/owner of a pressure system to select a competent person capable of carrying out the duties in a proper manner with sufficient expertise in the particular type of system, and we have established that in some cases, the necessary expertise will lie within the user's or owner's organisation. In such cases, the user or owner is acting as a competent person and is responsible for compliance with the Regulations. However, small or medium-sized businesses may not have sufficient in-house expertise and if this is the case, they should use a suitably qualified and experienced, independent, competent person. Whether the competent person is drawn from within their own organisation or from outside, they should establish that the person nominated has sufficient understanding of the systems to enable them to draw up schemes of examination or certify them as suitable.

Where the competent person is a direct employee of the user's or owner's organisation, there should be a suitable degree of independence from the operating functions of the company. In particular, where the staff are provided from an in-house inspection department and carry out functions in addition to their competent person duties, they should be separately accountable under their job descriptions for their activities as competent persons. They should act in an objective and professional manner with no conflict of interests and should give an impartial assessment of the nature and condition of the system.

A competent person capable of drawing up schemes of examination or examining a simple system may not have the expertise, knowledge and experience to act as a competent person for more complex systems. For a number of systems, including the larger or more complex, it is unlikely that one individual will have sufficient knowledge and expertise to act on their own and it is possible for a team of employees to be chosen as the "competent person" to provide the necessary breadth and depth of knowledge and experience.

The level of expertise required by the competent person depends on the size and complexity of the system in question.

STUDY QUESTIONS

4. Outline the elements of a strategy to prevent failure of pressure systems.
5. What basic information about pressure vessels should be permanently marked on the vessel?

(Suggested Answers are at the end.)

Summary

Work Equipment

We have:

- Explained why risk assessment must be carried out on work equipment (with reference to **PUWER** and **MHSWR**).
- Discussed the employers duty to ensure that all work equipment is fit for purpose (**PUWER**).
- Examined the means by which all forms of energy used or produced, and all substances used or produced can be supplied and/or removed in a safe manner.
- Considered ergonomic, anthropometric and human reliability in the use of work equipment, including the layout and operation of controls and emergency controls and reducing the need for access (including automation and remote systems).
- Examined the importance of the size of openings, height of barriers and distance from danger with regard to preventing access to dangerous parts of machinery/work equipment.
- Identified the range of risks associated with work equipment, specifically aspects of safety identified in **PUWER**:
  - Initial integrity.
  - Where it will be used.
  - Purpose for which it will be used.
  - Decommissioning and end of life.
- Discussed the risks associated with using work equipment which arise from its:
  - incorrect installation or reinstallation,
  - deterioration, or
  - exceptional circumstances,

  which could affect the safe operation of work equipment during maintenance, inspection and testing.
- Identified the hierarchy of controls for work equipment:
  - Remove all risk by design.
  - Eliminating the risks.
  - Taking 'hardware' (physical) measures (such as providing guards).
  - Taking appropriate 'Software' measures (such as following safe systems of work and providing information, instruction and training based on the level of risk associated with the equipment).

Pressure Systems

We have:

- Defined pressure systems.
- Considered the types of inspection, frequencies and statutory basis for examination of pressure systems.
- Examined prevention and testing strategies to prevent the failure of pressure systems, which should consider:
  - Design and construction.
  - Repair and modification.
  - Information and marking.
  - Safe operating limits.
  - Written scheme of examination.
  - Maintenance and record keeping.
  - The requirement for competent persons.

Unit ND3

Suggested Answers - Part 1

No Peeking!

Once you have worked your way through the study questions in this book, use the suggested answers on the following pages to find out where you went wrong (and what you got right), and as a resource to improve your knowledge and understanding.

Learning Outcome 11.1

Question 1

Under the **Health and Safety (Safety Signs and Signals) Regulations 1996**, an employer is required to provide safety signs whenever there is a residual risk which has not been avoided or controlled by other means.

Question 2

Prohibition sign - round with white background and red border and diagonal red bar.

Mandatory sign - round with blue background and white symbol.

Safe condition sign - square or oblong with white symbols on green background.

Warning sign - triangular black border with black pictogram on yellow background.

Question 3

When installing lighting at a workplace, employers should consider:

- The type of work being carried out.
- Levels of natural light available.
- Working conditions, e.g. flammable, dusty or explosive atmospheres.
- Legal requirements, i.e. **RRFSO** and **DSEAR**.
- Their duty to ensure the health and well-being of their workers and individual requirements.
- A robust system for the maintenance, replacement and disposal of lamps and luminaires.

Learning Outcome 11.2

Question 1

The specific risks associated with confined spaces are: fire and explosion, drowning, increase in body temperature, asphyxiation, and entrapment in a free-flowing solid.

Question 2

The atmosphere in a confined space can be harmful to workers due to: a lack of oxygen, build-up of toxic gas or build-up of flammable gas.

Question 3

A worker can enter or work in a confined space without BA if:

- There is a confined spaces permit to work in place.
- All electrical, mechanical, chemical, heat and other sources have been isolated.
- Effective steps have been taken to avoid ingress of dangerous fumes.
- Sludge or other deposits have been removed.
- The workplace has been cleaned, drained and purged, as necessary, for the type of work to be carried out and entry to be made.
- The space contains no material liable to give off fumes.
- The space has been adequately ventilated and tested for fumes.
- There is an adequate supply of respirable air.
- The space has been certified as being safe for entry without breathing apparatus.

Question 4

Equipment which should be available in case of an emergency includes: BA sets, resuscitators, means of summoning help (radios), life-lines, oxygen.

Learning Outcome 11.3

Question 1

- Flash point is the lowest temperature at which sufficient vapour is given off to flash, i.e. ignite momentarily, when a source of ignition is applied.
- Auto-ignition temperature is the lowest temperature at which the substance will ignite without the application of an external ignition source.

Question 2

With the flash point of petrol being so low (approx. –40°C), under normal conditions it will always be giving off sufficient vapour to be above the fire point. Any source of ignition is therefore liable to result in fire.

Question 3

Types of vapour cloud explosions include: Confined Vapour Cloud Explosion (CVCE), Unconfined Vapour Cloud Explosion (UVCE) and Boiling Liquid Expanding Vapour Explosion (BLEVE).

Question 4

A CVCE occurs if a flammable vapour cloud is ignited in a container, e.g. a process vessel, or in a building, so that it is confined. Pressure can build up until the containing walls rupture. A UVCE results from the release of a considerable quantity of flammable gas or vapour into the atmosphere, and its subsequent ignition.

Question 5

A primary explosion occurs in part of a plant, causing an air disturbance. The air disturbance disperses dust which causes a secondary explosion which is often more destructive than the primary explosion.

Question 6

When placed in a fire situation, unprotected steel will rapidly lose its designed shape and therefore its strength. In a fire, a steel beam will expand and push the columns out, causing the floor slabs to collapse onto the floor below.

This floor, not being strong enough to carry the extra load placed on it, also collapses. In a multi-storey building the effect is often that the whole building collapses floor by floor. The steelwork can also spread the heat by conduction, causing the fire to spread.

Question 7

Flame-retardant paint, when exposed to excessive heat, bubbles; this gives additional protection to the covered timber.

Question 8

The spread of fire within a building can be minimised by compartmentation and the use of suitable fire-resistant walls, floors and fire doors.

Question 9

Fire-stopping can be described as preventing the spread of smoke and flame by placing obstructions across air passageways. Ventilation ducts and gaps around doors must have the facility to be stopped in the event of a fire. This can be achieved by the use of baffles, self-closing doors and intumescent material which expands when subject to heat, thereby sealing the opening.

Question 10

Any four from the following features should be considered for buildings that could be at risk from dust explosions:

- Wherever possible, they should be isolated from other buildings.
- Buildings should preferably be one storey high.
- If inside, the vulnerable area of the building must be reinforced.
- The rest of the areas of the plant must be protected (e.g. a blast wall may be needed).
- Sufficient explosion venting to avoid structural damage from overpressure must be provided.
- Hot gases from the explosion must be vented to the outside atmosphere to prevent secondary fires.

Other features are safe escape routes in case of an explosion and fire, fire-resistant construction materials, fire-resistant doors and good electrical insulation.

Question 11

- Zone 0 is classified as a place in which an explosive atmosphere consisting of a mixture with air of dangerous substances in the form of gas, vapour or mist is present continuously or for long periods of time or frequently.
- Zone 20 is classified as a place in which an explosive atmosphere in the form of a cloud of combustible dust in air is present continuously, or for long periods or frequently.

Question 12

When considering explosion venting, consideration must be given to ensuring that any gas or liquid that may escape when a bursting disc/explosion panel fails are safely vented away from the operator.

Question 13

Bursting discs are the weakest point in the system - designed to fail and so avoid mechanical damage to the rest of the system.

Learning Outcome 11.4

Question 1

If an enforcing authority considers that premises constitute, or may constitute, a serious risk if changes were to be made to them or their use, it may serve an Alterations Notice on the responsible person. If such a notice is issued, the person receiving it must inform the enforcing authority of any proposed changes.

Where an enforcing authority is of the opinion that there has been a failure to comply with any provision of the **RRFSO**, it may serve an Enforcement Notice on the responsible person, or anyone who has control of the premises. The notice will indicate what steps may be taken to remedy the contravention and state a timescale for completion of the work involved.

The enforcing authority may serve a Prohibition Notice in any case where the risk to persons in case of fire is serious. The notice must:

- State that the fire authority is unhappy with the fire risk present.
- Specify the matters which in their opinion give rise to that risk.
- Direct that the use of the premises (or part of the premises) is prohibited or restricted until the risk is remedied. A Prohibition Notice may include steps to be taken to remove the risk from fire and will come into effect immediately.

Question 2

Under **RRFSO**:

(a) The responsible person in a workplace is the employer or any other person who may have control of any part of the premises, e.g. the occupier or owner. In all other premises the person or people in control of the premises will be responsible. If there is more than one responsible person in any type of premises, all must take all reasonable steps to work with each other.

(b) A relevant person is any person who is, or may be, lawfully on the premises, and anyone in the immediate vicinity who is at risk from a fire on the premises. This does not include fire-fighters when carrying out fire-fighting or other emergency duties.

Question 3

Duties of the responsible person under the **RRFSO**:

- Carry out a risk assessment.
- Take general fire precautions.
- Appoint one or more competent persons to assist in the implementation of preventive and protective measures.
- Implement appropriate arrangements for the effective planning, organisation, control, monitoring and review of preventive and protective measures.
- Apply the principles of prevention when implementing any preventive and protective measures.
- Provide employees with information on fire risks, preventive and protective measures and emergency procedures.
- Provide the employer of any other persons working on the premises with information on fire risks, preventive and protective measures and emergency procedures.
- Ensure that the premises, and any facilities or equipment provided in relation to fire safety, are maintained.
- Eliminate or reduce risks from dangerous substances.

- Ensure that appropriate equipment for detecting fire, raising the alarm and fighting fire is provided.
- Ensure that emergency routes and exits are kept clear at all times, lead directly to a place of safety, are adequate for the premises and the number of persons present, have appropriate doors that open in the direction of escape and are capable of being easily and immediately opened by any person in an emergency, are indicated by appropriate signs and are provided with adequate emergency lighting.
- Ensure employees are provided with adequate safety training.
- Establish appropriate procedures to be followed in the event of serious and imminent danger.
- Ensure that additional emergency measures in respect of dangerous substances are in place.

Question 4

The types of physical harm that could be caused to persons by a workplace fire include:

- Smoke inhalation causing burning to the lungs and triggering conditions such as asthma.
- Suffocation or respiratory difficulties caused by depletion of oxygen.
- Poisoning by inhalation of toxic gases given off by combustion products.
- Burning by heat, flames or explosion.
- Injury from falling or collapsing structures.
- Falls from a height while attempting to escape.
- Crushing injuries caused by panic or stampede.
- Injury from broken and flying glass.
- Mental or physical trauma.
- Death.

Question 5

A fire risk assessment should adopt a structured approach that ensures all fire hazards and associated risks are taken into account:

- Identify fire hazards - look for all the sources of ignition, fuel and oxygen that together might cause fire.
- Identify people at risk - anyone who may be affected, not just workers, e.g. maintenance staff, contractors, cleaners, security guards, etc. Also visitors, members of the public and those who may be at particular risk must be considered, e.g. young or inexperienced workers, people with mobility or sensory impairment, etc.
- Evaluate, remove, reduce and protect from risk - through the use of:
  - Preventive measures to remove or reduce the risk of fire breaking out (e.g. effective control of ignition sources, appropriate storage of flammable materials, good housekeeping, maintenance and inspection of equipment, etc.).
  - Protective measures to remove or reduce people being harmed in the event of a fire occurring (e.g. provision of automatic detection and alarm systems, adequate means of escape, fixed and portable fire-fighting equipment, etc.).
- Record, plan, inform, instruct and train - significant findings of the assessment should be recorded; emergency plans should be developed; appropriate information and instruction should be provided to relevant persons; appropriate training should be provided.
- Review - the fire risk assessment should be reviewed regularly to take account of such things as new or changed fire hazards; after a significant or major incident; after a significant change in the workplace, etc.

Question 6

The four types of detector commonly used in buildings are:

- Ionisation smoke detectors.
- Optical detectors.
- Radiation detectors.
- Heat detectors.

Question 7

Regular tests of an alarm system serve to check the circuits and to familiarise staff with the call note.

Question 8

Manual systems are suitable for small workplaces.

Question 9

The five classes of fire are:

- A - Solids (organic solids).
- B - Flammable liquids and liquefiable solids (subdivided into those miscible with water and those which are not).
- C - Gases and liquid gases.
- D - Flammable metals.
- F - Cooking fat fires.

Question 10

The colour coding of fire extinguishers and the types of fire that they are suitable for are as follows:

- Water - red - Class A fires.
- Foam - cream - Classes A and B fires.
- Dry powder - blue - Classes A, B and C fires.
- Dry powder - violet - Class D fires.
- Carbon dioxide - black - Class B and electrical fires.
- Wet chemical - yellow - Class F fires.

Question 11

'Starvation' refers to the removal of the fuel leg of the fire triangle. This can be achieved by taking the fuel away from the fire, taking the fire away from the fuel and/or reducing the quantity or bulk of fuel available. Materials may therefore be moved away from the fire (to a distance sufficient to ensure that they will not be ignited by any continuing radiant heat) or a gas supply may be turned off.

Question 12

Gas fires may be difficult to deal with because while dry powder and carbon dioxide may be used to knock the flame down, there is a risk of a build-up of gas if it cannot be turned off. In some situations, it may be preferable to evacuate the area, allow the fire to continue and call the fire service.

Question 13

The main factors to consider for an adequate means of escape are the:

- Nature of the occupants, e.g. mobility.
- Number of people attempting to escape.
- Distance they may have to travel to reach a place of safety.
- Size and extent of the 'place of safety'.

Question 14

A place of relative safety exists where a fire door is placed between people and the fire. Once within this place, the structure will protect people from smoke and heat while they make their way to a place clear of risk.

Question 15

Fire-resisting doors (smoke-stop) are used to:

- Break corridors into sections and thereby reduce the area of smoke logging.
- Separate stairs from the remainder of the floor area.
- Confine an outbreak of fire to its place of origin.
- Keep escape routes free of smoke for long enough to permit evacuation of occupants.

Question 16

Emergency (or safety) lighting should be provided where failure of the normal system would cause problems, e.g. in buildings used after dark, or darkened, e.g. cinemas, hospitals, and sections of buildings used for means of escape.

Emergency escape lighting should be provided in those parts of buildings where there is underground or windowless accommodation, core stairways or extensive internal corridors. Generally the need for such lighting will arise more frequently in shops than in factories and offices because of the greater likelihood of people in the building being unfamiliar with the means of escape.

Question 17

The following topics should be covered in each training session with practical exercises where possible:

- Action to take on discovering a fire.
- How to raise the alarm and the procedures this sets in motion.
- Action to be taken on hearing the fire alarm.
- Procedures for alerting members of the public including, where appropriate, directing them to exits.
- Arrangements for calling the fire service.
- Evacuation procedure for everyone in the premises to an assembly point at a place of safety.

- Location and use of fire-fighting equipment.
- Location of escape routes, including those not in regular use.
- How to open all escape doors.
- The importance of keeping fire doors closed.
- How to stop machines and processes and isolate power supplies where appropriate.
- The reason for not using lifts (other than those specifically provided or adapted for use by people with disabilities).
- The importance of general fire precautions and good housekeeping.
- The items listed in the **emergency plan**.
- The importance of fire doors and other basic fire-prevention measures.
- The importance of reporting to the assembly area.
- Exit routes and the operation of exit devices, including physically walking these routes.
- General matters such as smoking policy and permitted smoking areas or restrictions on cooking other than in designated areas.
- Assisting disabled persons where necessary.

Question 18

The **operation** of a fire marshal system involves:

- The building being split into small areas of responsibility.
- Each area is allocated to a specific fire marshal.
- Fire marshals are designated people who, in the event of a fire:
  - Search and check their allocated area.
  - Ensure that all people have left the building.
  - Direct those who have not left the building to an appropriate fire exit and safe assembly point.
  - Report that their area has been checked and is clear.
- The **benefits** of the system include:
  - The use of trained persons who are familiar with the premises to evacuate other people who may not be familiar with the premises.
  - Marshals can compensate for any adverse human behaviour which might hinder or delay the evacuation.

Question 19

To execute the Personal Emergency Evacuation Plan (PEEP) special aids may be needed such as:

- **Personal trembler alarms** which "vibrate" at the same time as the alarm.
- **A buddy system** where someone is allocated the task of helping the person with a sensory impairment.
- **Visual alarms** for the **hearing** impaired such as a flashing beacon.
- **Tactile/Braille** signs for the **visually** impaired, providing the person can locate the sign and is also able to read Braille.

Learning Outcome 11.5

Question 1

The rate of a reaction will increase exponentially with increase in temperature; in practical terms, an increase of 10°C roughly doubles the reaction rate in many cases.

Question 2

The two key factors in preventing thermal runaway reactions are mainly related to the control of reaction velocity and temperature within suitable limits.

Question 3

A suitable store for flammable liquids should have:

- Security to prevent unauthorised access, theft and arson.
- Clear signage at entrances to indicate the hazards of the contents and the key control measures such as no smoking and no naked flames.
- An impervious floor to the room that is compatible with IBC/drum contents.
- A sill to the room to create a bund capable of retaining 110% of the contents of the largest container.
- A slight slope to the floor so that leaks under IBCs/drums will flow out into open space where they will be visible.
- Fire-rated walls to keep fire out of the storage area for a period of time.
- Self-closing fire-rated doors to all entrances/exits.
- Fire-rated glass to any windows.
- A lightweight roof that will lift off easily in the event of a confined vapour cloud explosion in the building (so preventing overpressure from blowing out the walls).
- Good levels of natural ventilation through low-level and high-level vent/louvres.
- Segregation of incompatible materials.

Question 4

The main means of spillage containment may be an impervious sill or low bund, typically 150mm high, and big enough to hold 110% of the contents of the largest container.

Question 5

To avoid overfilling, care should be taken to ensure that the receiving container is large enough to receive the total amount of substance that is going to be delivered. The process must be supervised throughout the filling and delivery process.

Question 6

Spray booths are typically classified as Zone 1 areas - where an explosive gas/air mixture is likely to be present during normal operation.

Question 7

When selecting a container for filling, the container being used should be **suitable** for the substance that is being put into it. It must also be undamaged with no signs of leakage, staining or corrosion.

Question 8

Dispensing of flammable liquids should be done in a well-ventilated area in order to quickly dilute any escaping vapour concentrations to below the lower flammable/explosion limit.

Question 9

Unloading of a substance from a tanker into the wrong storage tank can be prevented by strict operating procedures and the use of couplings of a different design for each substance.

Question 10

Equipment which is liable to damage by corrosion due to the presence of moisture should be totally enclosed in an enclosure built and tested to the relevant Index of Protection (IP) standard such as IP66.

Question 11

Flameproof equipment: parts which can ignite a potentially explosive atmosphere are surrounded by an enclosure which can withstand the pressure of an explosive mixture exploding inside of it and prevents the propagation of the explosion to the atmosphere surrounding the enclosure. It can be used in Zones 1 or 2.

Question 12

Purging is the process of flushing hazardous gas or dust out of an enclosure prior to the introduction of the inert gas.

Question 13

Zone 2 classification is given to an area in which an explosive atmosphere consisting of a mixture with air of dangerous substances in the form of gas, vapour or mist is not likely to occur in normal operation but, if it does occur, will persist for a short period only.

Question 14

An emergency plan for dealing with victims of an explosion in a factory manufacturing paints and varnishes would have to consider the following:

- Treatment of the injured people on site.
- Transportation of the injured people to hospital.
- Evacuation of the building and site.
- Containment of the fire (to prevent further damage/injury from fumes, etc.).

The emergency plan would outline the actions required immediately after the explosion. For example, these could include:

- Raise the alarm (contact emergency services, e.g. fire and rescue service, police and ambulance).
- Evacuate the building/site.
- Advise the local community to keep windows closed, etc.
- Consider evacuation of local houses.
- Seal off other supplies to prevent further explosions/spread of fire.

Consideration with respect to dealing with the victims of the explosion would include ensuring that adequate first-aid facilities and first aiders are available.

Once the site has been made safe and the victims have been looked after, the company then has to start considering its recovery plan.

Details of the emergency plan should be shared with the local emergency services and other relevant external bodies such as the water company and Environment Agency. Information relating to the types of chemicals stored, and any fumes likely to come off as a result of fire, explosion or spillage, should be identified.

Question 15

(a) The emergency evacuation procedures for each site should be developed by considering the types of hazards that are present, e.g. dangerous substances, and the types of emergency situations that could occur, e.g. fire, explosion, flood, etc. Consideration must also be given to whether an emergency on site could affect neighbouring houses or the environment. Once these have been determined, safe routes for evacuating the buildings at each site must be identified ensuring that the routes give adequate protection against the types of emergency that could potentially occur. Additionally, a safe assembly point must be identified to where people on site must evacuate. The names of people on site authorised to set emergency procedures in motion, and the people authorised to take charge of and co-ordinate the on-site mitigatory action, must be identified. All this would be completed with liaison with the local emergency services and external bodies, where appropriate.

(b) All employees must be trained in the emergency procedures and this will be a part of the induction training. In addition, refresher training will be carried out at regular intervals (e.g. annually). Written notices of the procedures will be displayed at appropriate points throughout the workplace. Where employees work at more than one site then they will receive a briefing for each site.

(c) The effectiveness of the procedures will be monitored by having practice evacuations of each site and by questioning employees as to what they understand the procedures to be. This questioning could be included in a health and safety audit.

Learning Outcome 11.6

Question 1

(a) Advantages of the operation of a planned preventive maintenance system:

- Safer operation.
- Reduced downtime and lost production.
- Increased reliability of equipment.
- Failure-rate predictions.
- Rapid turnaround of parts.
- Maintenance carried out at times of least disruption.
- Reduced risk from planned maintenance operations.

(b) Disadvantages of the operation of a planned preventive maintenance system:

- Over-maintenance, with reduced efficiency.
- High costs of parts still with useful life.
- Management time for planning and operating schedules.
- Over-familiarity with tasks leading to complacency.
- Higher skill level.
- Increased storage requirements for spare parts.

Question 2

Factors to be considered in developing a planned maintenance programme for safety-critical components include:

- Importance in the process.
- Machine complexity.
- Relationship with other machines.
- Availability of replacement equipment.
- Identification of critical components.
- Environmental factors.
- Maintenance capability within the company.
- Maintainability within the design of the machine.
- Unknown factors.

Question 3

The factors to be considered when determining inspection regimes for work equipment include:

- The type of equipment:
  - Are there safety-related parts which are necessary for safe operation of the equipment, e.g. overload warning devices and limit switches?
- Where it is used:
  - Is the equipment used in a hostile environment where hot, cold, wet or corrosive conditions may accelerate deterioration?
- How it is used:
  - The nature, frequency and duration of use will determine the extent of wear and tear and the likelihood of deterioration. In addition, equipment regularly dismantled and re-assembled will require inspection to ensure that it has been installed correctly and is safe to operate.

Question 4

The use of ultrasound as a non-destructive testing method involves the use of a generator transmitting ultrasound waves into a material and detecting them when reflected from within the material. It is analogous to the geological mapping process that uses reflected shock waves to interpret discontinuity phenomena underground.

The ultrasound waves are generated in a head that is placed upon and moved across the surface, generally with some form of lubricant (e.g. water) to minimise any air gap. The ultrasound travels into the material and is 'bounced' back to a receiver mounted in the head. The output is read on an oscilloscope. The equipment must be calibrated to the full depth of the material before starting. Any defect will cause a variation in the return signal and this can be interpreted to indicate the depth of the defect, so it can detect defects within the material that do not show on the surface.

Question 5

As an example, if you have chosen radiography and dye penetrant, the advantages and disadvantages might look like this:

Test	Advantages	Disadvantages
Radiography	Permanent, pictorial, easily interpreted images obtained. Locates majority of internal defects.	Radiation safety hazards. Expensive X-ray sets. Thickness limits (more so with X-rays). Power supply required.
Dye Penetrant	Cheap and convenient. Superior to visual examination. For all non-porous materials.	Surface defects only. Defects must be open to the surface.

Question 6

Control measures for maintenance work on running or working equipment:

- Providing temporary guards.
- Limited movement controls.
- Crawl speed operated by hold-to-run controls.
- Personal protective equipment.
- Provision of instruction and supervision.

Question 7

(a) Energy sources may be in the following form:

- Electrical.
- Mechanical.
- Hydraulic.
- Pneumatic.
- Chemical.
- Thermal.
- Gravitational.
- Radiation.
- Stored or kinetic energy.

(b) Isolation measures include:

- Locks.
- Clasps.
- Tags.
- Closing and blanking devices.
- Removal of mechanical linkages.
- Blocks.
- Slings.
- Removal from service.

Learning Outcome 11.7

Question 1

The factors that need to be considered when identifying the risks associated with the use of work equipment include:

- Initial integrity.
- The location where it will be used.
- The purpose for which it will be used.
- Incorrect installation or re-installation.
- Deterioration.
- Other exceptional circumstances which could affect its safe operation.

Question 2

A significant risk is one which could foreseeably result in a major injury or worse arising:

- from incorrect installation or re-installation;
- from deterioration; or
- as a result of exceptional circumstances which could affect the safe operation of the work equipment.

Question 3

The three elements of the risk control strategy relating to work equipment are:

- Eliminate the risks created by the use of work equipment by careful selection.
- Employ 'hardware' (physical) measures such as:
  - Suitable guards.
  - Protection devices.
  - Markings and warning devices.
  - System control devices, such as emergency stop buttons.
  - Personal protective equipment.
- Employ 'software' measures such as:
  - Following safe systems of work.
  - Ensuring maintenance is only performed when equipment is shut down.
  - Provision of information, instruction and training.

Question 4

The key points in a prevention strategy are:

- Fit legally-required safety devices.
- Operate the system within the limits specified.
- Undertake the statutory inspection of the full system.
- Undertake more regular inspections between statutory inspections.
- Use Non-Destructive Testing (NDT) techniques to determine what is happening inside the system.

Question 5

The information required to be permanently marked on pressure vessels is:

- The manufacturer's name.
- A serial number.
- Date of manufacture.
- The standard to which the vessel was built.
- Maximum and minimum allowable pressure.
- Design temperature.